# The Violence of Climate Change

Also by Kevin J. O'Brien

*An Ethics of Biodiversity: Christianity, Ecology, and the Variety of Life*

# THE VIOLENCE OF CLIMATE CHANGE

## LESSONS OF RESISTANCE FROM NONVIOLENT ACTIVISTS

### KEVIN J. O'BRIEN

Georgetown University Press
Washington, DC

The publisher is not responsible for third-party websites or their content. URL links were active at the time of publication.

Library of Congress Cataloging-in-Publication Data

Names: O'Brien, Kevin, 1977– author.
Title: The violence of climate change : lessons of resistance from nonviolent activists / Kevin J. O'Brien.
Description: Washington, DC : Georgetown University Press, 2017. | Includes bibliographical references and index.
Identifiers: LCCN 2016033266 (print) | LCCN 2016040147 (ebook) | ISBN 9781626164345 (hc : alk. paper) | ISBN 9781626164352 (pb : alk. paper) | ISBN 9781626164369 (eb)
Subjects: LCSH: Climatic changes—Effect of human beings on. | Climatic changes—Social aspects. | Global warming.
Classification: LCC QC903 .O36 2017 (print) | LCC QC903 (ebook) | DDC 363.738/74—dc23
LC record available at https://lccn.loc.gov/2016033266

♾ This book is printed on acid-free paper meeting the requirements of the American National Standard for Permanence in Paper for Printed Library Materials.

18 17      9 8 7 6 5 4 3 2   First printing

Printed in the United States of America

Cover design by Martyn Schmoll. Cover image by iStock.com/piyaset.

# CONTENTS

# ACKNOWLEDGMENTS

First, thanks to five people I have never met—Cesar Chavez, Martin Luther King Jr., Dorothy Day, Jane Addams, and John Woolman. This book seeks to pay tribute to their lives, their thoughts, and their witnesses. I hope that in a small way it honors them and all who carry on their legacies.

Learning ethics from historical figures would have been beyond my capacity without the lessons of Michael Birkel, my very first professor of religion. He taught me how to study people from the past with careful, scholarly attention to the wisdom they offer about living well in the world. I do not do it nearly as well as he does—and all the failings of this book are mine alone—but I keep trying, and it is my great joy that Michael has remained a mentor and friend for almost twenty years now. I thank him for that, and for his most recent gift, a generous and thoughtful reading, analysis, and critique of chapter 3.

Deep thanks also to Megan Noborikawa, who taught me a lot when she was my student and then far more when she did a comprehensive edit of an early draft of this book. She helped to clarify both the argument and how it was expressed. Having such a smart and diligent former student gives me hope for the future.

Thanks to Dan Spencer and Willis Jenkins for generous and critical readings. Both provided assessments of this manuscript for Georgetown University Press, helping to expand its audience and shape a far better book. They began this work as anonymous peer reviewers, but each subsequently revealed his identity and continued constructive feedback in conversation.

Many other colleagues have been long-standing collaborators and interlocutors, and I thank all the scholars who work on religious ethics and environmental issues. Most particularly, I offer my gratitude to Jennifer Ayres, Whitney Bauman, Trevor Bechtel, Kate Blanchard, Rick Bohannon,

Forrest Clingerman, Sarah Fredericks, Laura Hartman, and Darryl Stephens. I do not know how I would think or write without these friends.

My colleagues at Pacific Lutheran University also offer continuing support and expertise, helping to make this school a place to believe in as well as to work. I offer particular gratitude to Agnes Choi, Suzanne Crawford O'Brien, Seth Dowland, Tony Finitsis, Aimee Hamilton, Erik Hammerstrom, Brenda Llewellyn Ihssen, Brian Naasz, Doug Oakman, Jason Skipper, Jennifer Smith, Claire Todd, Samuel Torvend, Marit Trelstad, Michael Zbaraschuk, and Joel Zylstra.

I could not work without a range of conversation partners and inspirations outside academia. I have been privileged to volunteer for the last ten years with Earth Ministry, which draws on religious faith to nurture positive environmental change. Thanks especially to its current and former staff members LeeAnne Beres, Jessie Dye, Jessica Zimmerle, and Clare Josef-Maier. I am also grateful to two Christian peace communities—Agape in western Massachusetts and Corrymeela in Northern Ireland. Years ago, I was lucky enough to live briefly in each community, and they have had an impact on my life ever since.

Thanks to Richard Brown of Georgetown University Press for his help imagining this project over five years and then for encouraging me through its completion. I also thank the book's copyeditor, Alfred Imhoff, its project editor, Kathryn Owens, and the rest of the staff at the press whose efficiency and intelligence made the process a pleasure and made the book so much better.

Finally, thanks to my family for a lifetime of unconditional support, nurturing, teaching, and challenge. My nieces, nephews, brothers, parents, and extended family made me who I am and keep me striving toward who I want to be. My gratitude is furthest beyond words for two of these family members. My wife, Mary, is the best partner I could ever hope for. Thank you, Mary, for making life better and richer every day. My mother, Mary Lou McCloskey, taught me what it means to live fully and for others and to try to leave every part of the world better than I find it. I dedicate this book to you, Mom, with thanks for all you have done and all that you are.

# Introduction

## Toward a Witness of Resistance

Introducing his novel *Slaughterhouse-Five*, Kurt Vonnegut recounts a conversation in which someone scoffed at the idea of an "anti-war book," asking him, "Why don't you write an anti-*glacier* book instead?" Vonnegut follows with a straightforward interpretation: "What he meant, of course was that there would always be wars, that they were as easy to stop as glaciers."[1]

In the 1960s Vonnegut and his critic both assumed that glaciers were inexorable forces of nature beyond human influence. In a world increasingly shaped by climate change, we know that this is not true. Near where I write these words, the glaciers of Mount Rainier shrank by 25 percent in the twentieth century and are shrinking even more rapidly in the twenty-first. This means a less predictable water supply for millions of people, raising doubts for the region's growing population.[2] A more dramatic case of glacial decline is occurring in Antarctica, where—the National Aeronautics and Space Administration predicts—the West Antarctic Ice Sheet, a continental mass that feeds numerous glaciers, will likely disappear in coming centuries because of warming global temperatures. This will result in rising sea levels all over the world, posing a profound long-term threat to coastal communities everywhere.[3]

This book is about climate change, inspired partly by the fact that glaciers—symbols of constancy and inevitability in Vonnegut's conversation—are now rapidly shrinking because of industrial human activity.[4] This is also a book about war, asking what concerned people in the twenty-first century can learn from the examples of five particular Christians in the United States who responded to the violence of the eighteenth, nineteenth, and twentieth centuries with nonviolent resistance. Using this tradition of

nonviolence as an inspiration for a response to climate change, this book
seeks to stand up for glaciers and against violence.

## CLIMATE CHANGE AS VIOLENCE

My premise is that climate change is a problem of structural violence.
Atmospheric changes are structural because they are the result of countless small decisions and developments in politics, economics, and technology. Climate change is not caused by any one person's decision, and
no individual can stop it. The structural nature of this problem too often
makes it seem invisible, an abstraction that public discourse suggests may
or may not be real, may or may not be caused by human beings, and may
or may not be related to the latest extreme weather events. But climate
change is very real, very much caused by human beings, and very much
connected to hurricanes, droughts, and floods.[5]

These structural changes to the atmosphere are violent because they
are harming life on Earth and its future. Species die off, glaciers disappear, and forests dry up. The ecosystems upon which future generations of
humanity depend are endangered. Today, refugees flee their homes as sea
levels rise, people fight over ever-scarcer water supplies, and farmers work
ever harder to feed their families, their communities, and the world. This
is violence, a product of human actions that hurts others.

Climate change has been created by generations of decisions from
privileged people who seek to make themselves safe and comfortable,
who contribute disproportionately to the problem of climate change while
tending to avoid its worst effects. This is well demonstrated by my own
community: economically secure persons in the United States. The average US citizen is responsible for emitting about 17 tons of carbon dioxide
and other climate-changing gases each year; the average global citizen is
responsible for about 5 tons.[6] But even in a country like the United States,
the per capita emissions of the very poor tend to be well under 5 tons
each year, while the very wealthiest are responsible for upward of 70 tons a
year.[7] Climate change is mostly caused by wealthy people.

I hope that many different audiences can learn from this book, but
every reference to "we" or "our" or "concerned people" is directed at privileged people in the industrialized world with relatively extensive resources,

who bear guilt for and a responsibility to oppose the violence of climate change. As a member of this group, I am writing about a personal struggle. I drive a car most days, I fly on planes a half dozen times a year, and I eat food shipped from around the world. All these activities release climate-changing gases into the atmosphere. I write these words on a computer with both a smartphone and tablet nearby; and these are just a few examples of the technologies that support my lifestyle and require substantial resources for their construction, use, and disposal. I am part of, complicit in, and dependent upon the systems that cause climate change.

And yet, in a dark irony, climate change does not threaten me in the ways it threatens others. I live near but well above the ocean, in a city prosperous enough to ensure that people like me will have regular access to clean and abundant water and energy even as melting glaciers change the flow of rivers that support life and provide hydroelectric power. I am a white, middle-class citizen of a nation that has proven it will go to great lengths to rescue middle-class and wealthy white people who suffer from natural disasters. I live in a fertile place that is likely to remain fertile for the foreseeable future, even as weather patterns shift. Of course, all of us are threatened by the changing climate, but it is unjust that I am less threatened than many others around the world.

While teaching and writing about environmental issues for the last decade, I have become convinced that the degradation of the planet's ecosystem is best understood as violence and that I am guilty of this violence. I have written this book in order to wrestle with this troubling fact. But I share the book in hopes that this struggle will be not merely personal—that the book might help others who are similarly troubled and that it might spur some people who are not yet troubled to become so.

All living beings are affected by climate change. This book is written for those who, like me, are causing this suffering through our everyday lives and who want to do something about it. Therefore, it argues that concerned people must face the violence of climate change with the resources of nonviolence.

The definition of nonviolence used in this book has two dimensions: Nonviolence is (1) a commitment to actively oppose violence (2) without the use of violence. The emphasis through most of the book is on the first dimension, not because the second is unimportant but because it is less relevant to contemporary struggles against climate change. Few privileged people have been tempted to use violence in response to a changing climate. Even those willing to break the law for this problem have made great efforts to do so without endangering human lives. Because there is little chance of widespread violent activism against climate change, this book does not focus on the importance of avoiding violence in the course of our protests. Rather, the central claim here is that the tradition of nonviolence calls for creative and active resistance against the violence of climate change, that it is time to act in an organized, thoughtful, and faithful way. The problem facing people of privilege in a world of climate change is the temptation to do nothing—to deny violence, ignore it, or trust that others will solve it.

To see climate change as violence is to see it as the product of a destructive system that degrades human lives, other species, and the world upon which all living beings depend. To live nonviolently is to seek ways to engage in political debate, shape culture, feed oneself, build transportation systems, and gather energy without causing suffering or harm. This is a wildly idealistic goal, but the chapters that follow demonstrate that some of the most effective and influential people in US history have been wildly idealistic. The best way to resist the systems of violence that grip US society and the world is to imagine a future of justice and then pull the world in that direction.

## ETHICAL COMMITMENTS: SEEKING JUSTICE IN CONVERSATION

This book argues that privileged people in the twenty-first century are called to oppose the violence of climate change and that we can find resources for doing so in nonviolent social movements. This argument is informed and guided by four commitments: (1) to seek climate justice, (2) to learn from social movements, (3) to draw from Christian exemplars as a contribution to an inclusive conversation, and (4) to combine abstract and concrete thinking in hopes of arriving at creative responses to a complex problem.

### The Goal of Climate Justice

The first commitment is to seek the most just response to climate change rather than a solution to it. Climate change is a reality with which the human race will be living for many generations. As is further explained in the next chapter, the atmosphere has fundamentally changed because of human activity, and these changes are already harming human beings. It is no longer possible to talk about preventing or ending climate change.

In response, the environmental activist Tim DeChristopher writes that concerned people are called to build "a movement for climate justice" that seeks "to defend the right of all people, and not only people of all races and nationalities but people of all generations, to live healthy lives and have both the agency and the environment necessary to create the lives they want. We are building a movement to hold onto the things about our civilization that are worth keeping. We are building a movement to navigate that period of intense change in a way that maintains our humanity."[8] Climate justice means accepting the fact that climate change is a reality but refusing to accept the mistakes that created it or the inequities and violence it causes.[9]

This book seeks to contribute to the movement for climate justice by bringing into the conversation witnesses from the past who maintained humanity in the face of grave injustice. It also seeks to extend the witness of such people by considering how their ideas are relevant to the very real and immediate challenges facing humanity in the twenty-first century, hoping to extend the goal of justice beyond even the human species, to all creatures.

Climate justice calls us to understand environmental issues as fundamentally linked to other moral challenges, such as racism, sexism, and economic injustice. As the ethicist James Nash writes: "There can be no social justice without ecological justice! There can be no peace among nations in the absence of peace with nature!"[10] This book seeks to show that the movement for climate justice is, ultimately, a movement for peace between human beings, human communities, and all other creatures.

## Learning from Diverse Social Movements

The second commitment is to be led as much as possible by the voices of the poor, the marginalized, and the oppressed and by those who have sought to resist poverty, marginalization, and oppression. Too much of the moral discussion about climate change has been what the theologian James Cone calls a "monological" conversation—"a dominant group talking to itself." Privileged white people have drawn on one another's ideas to ponder the mistakes made by privileged white people. Cone notes that it should be clear by now that such monological conversations do not pay adequate attention to either the Earth's systems or the poor and marginalized peoples who depend most immediately upon these systems. Cone calls on privileged people to "look outside of their dominating culture for ethical and cultural resources for the Earth's salvation."[11]

In 2004 the ethicist Larry Rasmussen echoed this call, suggesting a basic reassessment of the methods, hopes, and goals of ethics. He borrowed a line from James Baldwin in asking scholars to "do our first works over," reassessing and relearning what it means to think ethically in light of the environmental justice movement. This movement includes and advocates for people whom environmental degradation makes sick, thirsty, and homeless, who share "the collective experience of injustice."[12] If ethicists want to seek a more sustainable future, they should learn first and foremost from those who have sustained communities in the face of systems designed to destroy them, such as poor people and people of color. If concerned people seek to extend moral consideration beyond the merely human world, we have much to learn from those communities that have been dismissed as less than human in the past, including women and indigenous communities.

Rasmussen modeled and suggested a discourse with contemporary environmental justice activists. This book extends his suggestion temporally backward, attempting to learn about justice from movements of the eighteenth, nineteenth, and twentieth centuries. Two of the five witnesses whose lives are explored in this book were persons of color who struggled against racism, two were women who struggled against sexism, and all were willing to "do their first works over" when they encountered violent injustice.

## Christianity as Conversation Partner

The third commitment is to draw inspiration and guidance from Christians, but to do so in a way open to inclusive conversation with people of diverse faiths and no faith. As a Christian trained in the academic field of Christian ecological ethics, I am particularly interested in moral ideas emerging from my religious tradition.[13] However, I do not intend to suggest that this tradition is universally or unquestionably an exemplar of environmental and social virtue. Many Christians have done terrible things, have had destructive ideas, and continue to exacerbate injustices and environmental destruction. For this reason, I make no attempt to apologize for, defend, or even learn from "the Christian tradition" as a whole. It is too broad, too diverse, and too flawed to teach anything in its singularity. Instead, I focus on a particular line of thought within Christianity—nonviolent resistance—and seek to learn from five exemplars. This is not a book about a generic "Christian" response to climate change, but instead an argument that a particular aspect of Christianity has something to teach about climate justice.

Much Christian ecological ethics has been written primarily for Christians, advocating a particular kind of faith, worship, and community in response to environmental problems. For example, Laura Yordy argues that churches are called to be a "witnessing body to the Kingdom of Heaven which, as promised by God, encompasses the redemption of all God's creation." For Yordy, such "witness" is for Christians, and this allows her to make a very specific argument that all Christians should dedicate their lives to worshipping God, whom she asserts has already redeemed the world and thus calls Christians to live at peace with creation.[14]

The witness called for in this book is different because it is directed not at Christians but at all privileged people, who are called to resist the violence of climate change whether they share a belief in God or not. This book is for everyone concerned about climate change, everyone open to learning from the lives and works of five Christians who resisted violence. Although I make an argument in chapter 5 that the movement for climate justice should take faith and religion seriously, I neither assert nor assume that everyone should share a common faith or adhere to a single religious tradition. I understand the movement for climate justice as a broad and diverse phenomenon, and though I argue that Christian voices can inform this movement, I see them as part of a much broader conversation.

## Synthesizing the Abstract and the Concrete

The final methodological commitment of this book is for moral attention to both concrete action and abstract cosmological worldviews. Most
environmental ethics to date has looked to religion primarily for broad
cosmological vision. For example, in her book *A New Climate for Theology*, Sallie McFague calls for religious ethics because it determines people's
"deeply held and often largely unconscious assumptions about *who we are
in the scheme of things and how we should act*" (emphasis in the original).[15]
She presents the moral response to climate change as a theological choice
between the hierarchical, individualistic worldview of neoclassical economics and the integrated, communal worldview of ecological economics.
This is a vitally important cognitive shift; but it is incomplete on its own.

The ethicist Willis Jenkins calls for a counterbalance to such abstract
ethics, advocating a pragmatic approach that begins not by rethinking
basic assumptions but rather with "concrete problems and doing ethics
with imperfect concepts and incompetent communities." In other words,
ethics should not be about determining the right view of the world and
then applying it to problems; rather, it should be about wrestling with real
problems in conversation with broad claims about the nature of reality.
From this perspective, ethics creates new ways of thinking but only in dialogue with communities whose members seek to "create new possibilities
from their inherited traditions."[16]

Jenkins calls for this approach in direct response to climate change,
which he characterizes as an "unprecedented" moral challenge.[17] Given
the complexity of this problem, he argues, it is vital to learn from concrete action as well as broad worldviews. Whereas McFague proposes a
new way of thinking that will undo the damage of climate change, Jenkins
argues that the only realistic way forward is accepting the reality of climate
change and learning from those who are dealing with it day by day. This
is not a rejection of abstract thought—careful examination of big ideas
remains essential—but rather an insistence that concrete and grounded
analysis is a vital complement and counterbalance to broad claims about
how the world works.

Thus, this book begins not with an abstract claim but with a concrete
challenge, seeking first to properly understand climate change and then to

learn from inherited traditions about how to engage it realistically while striving to act on our highest ideals.

## A CLOUD OF WITNESSES

This book's argument for climate justice seeks to balance the practical and the ideal in conversation with social movements, using the Christian tradition to shed light on the contemporary challenge of climate justice. Its primary sources are the lives and writings of five nonviolent activists from the history of Christianity in the United States: John Woolman, Jane Addams, Dorothy Day, Martin Luther King Jr., and Cesar Chavez.

A community is strongest when it understands its traditions. The earliest Christians helpfully expressed this idea in the biblical Letter to the Hebrews, which recounts the stories of Noah, Abraham, and Moses to remind its readers that they have generations of Jewish history from which to learn. The letter then offers an inspirational call to action: "Since we are surrounded by so great a cloud of witnesses, let us also lay aside every weight and the sin that clings so closely, and let us run with perseverance the race that is set before us" (12:1).[18] This book turns to the tradition of nonviolent activism in order to summon a cloud of witnesses that will help twenty-first-century people to face climate change with perseverance, to run the long and challenging race of making the world more just.[19]

John Woolman (1720–72), a Quaker abolitionist, offers a lesson about the importance of transforming oneself. He sought not only to make the moral case against slavery but also to cleanse his life of all the privileges that slavery afforded to white men, leading him to give up his business and refuse to dye his clothes. His attempt to purify himself is an example to concerned people in the twenty-first century, who must decide how to deal with our own complicity in climate change.

Jane Addams (1860–1935) was a reformer and social worker who inspired national and global engagement as she built Chicago's Hull House. Her witness teaches a lesson about the scales of activism. Serving the poor and immigrant population of her neighborhood throughout her life, she had a direct impact on the people all around her, but she also traveled the world to advocate global peace and lobbied the US government to create a social safety net. Those who are tempted to rush to conclusions without carefully considering the diversity and complexity of climate change have much to learn from her work, which stressed the importance of mutuality and common ground across local, national, and global scales.

Dorothy Day (1887–1980), an activist and author who helped to found and to run the Catholic Worker movement, demonstrates the power of faith and love in activist work. Because of her deep religious commitments, she chose voluntary poverty and dedicated her life to the poor, and she repeatedly insisted that she could only remain faithful in community with others who shared her beliefs and who would regularly remind her of their implications. Her example calls twenty-first-century activists to consider the various ways that both organized and individual faith commitments can fuel the struggle for climate justice.

Martin Luther King Jr. (1929–68), a Baptist preacher who became the most famous voice in the civil rights movement, is a model of hope in troubling times. King stood up against racism and endured the frustrations of long protests because he trusted that the universe is ultimately on the side of justice and peace. Avoiding both cynical despair and blind optimism, his hope is a vital example for anyone who works toward justice in a world of uncertainties.

Cesar Chavez (1927–93), a union organizer who led a nonviolent campaign of farmworkers for more than thirty years, offers a lesson about the strategic use of sacrifice. Chavez disciplined himself by taking low wages and frequently fasting throughout his life, securing moral authority with the liberating power that he found in self-surrender. He also demanded sacrifices of his movement, recruiting volunteers rather than salaried employees to run his union and insisting that members contribute dues and go on strike in spite of financial hardships. These moves helped to create a unified community. Chavez has much to teach privileged people who recognize that climate change calls for both personal and political sacrifices.

In a 1954 speech, Martin Luther King Jr. spoke about the importance of individuals "who have the insight to look beyond the inadequacies of the old order and see the necessity for the new. These are the persons with a sort of divine discontent."[20] King himself was a person of divine discontent, as were Chavez, Day, Addams, and Woolman. These five witnesses offer resources for those who are discontented with the violence of climate change and who want to work toward a more just world where people can live in harmony with one another and all other creatures.

Of course, the five witnesses discussed in this book do not entirely capture the tradition of nonviolent protest in the United States, and none by herself or himself encompassed an entire movement. All were supported and challenged by colleagues and communities, and they are only one small part of the cloud of witnesses available to contemporary activists. However, these witnesses each played a key role in an important movement, and the life of each one offers an exemplary record of struggle for those of us who resist violence today.

These witnesses were human beings, and as such they were imperfect. This is best demonstrated by those who lived most recently: Chavez was a flawed leader who was not always able to hear dissent and seemed uncomfortable distinguishing the movement from his own personality.[21] King was also limited, and the intense scrutiny he received throughout and after his life revealed a series of adulterous affairs and a pattern of plagiarism in both academic and public writing.[22] Although their lives are less fully chronicled, Day, Addams, and Woolman were also fully human, and so flawed. No one becomes a witness for justice and peace by being perfect. Rather, we can learn from these real people who struggled to be good.

None of these witnesses was concerned about climate change, and none offers an answer to it. Only Chavez would ever have even heard of this problem, and while he was active in protesting environmental degradation and social injustice, he had nothing to say about the climate. The value of these witnesses is not that they offer answers to the present challenge but instead that they model a thoughtful moral response to other cases of structural violence. They offer guides—but not a plan—for a twenty-first-century movement seeking climate justice.

The treatment each figure receives in this book is also limited. Each one deserves and has had many books devoted to analyzing their writings, their lives, and the communities that made both possible. However, each has only a chapter here, which will, I hope, inspire further investigation into their work alongside further engagement with the work of resisting violence.

## THE PLAN OF THE BOOK

Before turning to each witness, chapter 1 offers an introduction to some of the basic facts about climate change and argues that it is helpfully understood as a "wicked" problem—that is, a problem that is multifaceted and has no clear solution—of structural violence. Chapter 2 then defines nonviolence by introducing the broad tradition and some of its expressions in Christianity.

Chapters 3 through 7 focus in turn on each of the five witnesses, who are covered in chronological order to demonstrate the developing legacy of nonviolence in the United States. In each case, the focus is on the witness's own writings. Each chapter draws a broad lesson about one particular topic relevant to climate justice—self-purification, the scale of action, faithful love as motivation, hope for the future, and sacrifice for climate justice. In addition, each chapter applies this lesson to a more concrete issue: personal austerity, balancing the rhetoric of social justice against the good of all species, considering the role of religion in the climate justice movement, assessing proposals to technologically engineer the climate, and judging demands that the industrialized world owes a "climate debt" to the global poor.

The conclusion to this book then draws on the common themes from these discussions to argue that privileged peoples must resist the violence of climate change even as we recognize how far our current practices, worldviews, and structures are from the ideal of climate justice. Recognizing that this ideal is distant from today's reality gives us all the more reason to push toward it and to learn from the cloud of witnesses who have done such work in the past.

## NOTES

1. Vonnegut, *Slaughterhouse-Five*, 3–4.
2. "Mt. Rainier National Park," www.glaciers.pdx.edu/Projects/LearnAboutGlaciers/MRNP/Chg00.html.
3. "The 'Unstable' Antarctic Ice Sheet: A Primer," National Aeronautics and Space Administration, www.nasa.gov/jpl/news/antarctic-ice-sheet-20140512/#.U5B5CRaE5tc.

4. Vonnegut identified the theme of *Slaughterhouse-Five* as less about opposition to war than about "the inhumanity of man's inventions to man," a phrase that is profoundly relevant to the violence of climate change. Vonnegut, *Slaughterhouse-Five*, xiii.

5. This book is not about the denial of climate change; it is directed to those concerned about the problem rather than those seeking to dismiss it. However, the basic premise that climate change is a problem of structural violence sheds light on the challenge of denial. To deny the reality of climate change despite scientific evidence and humane experience of degradation is to avoid a structural reality; to deny climate change is to deny that the systems in which we live and upon which we depend are contributing to violence and destruction. So, arguing against a climate change denier requires discussion not only of natural science but also of complicated social and cultural systems.

6. Carbon Dioxide Information Analysis Center, "Fossil-Fuel $CO_2$ Emissions," http://cdiac.ornl.gov/trends/emis/meth_reg.html.

7. Gough et al., "Greenhouse Gas Emissions." These numbers come from a study of emissions in the United Kingdom, which is not exactly comparable but offers clarity about the importance of economic status in determining per capita emissions.

8. Quoted by Stephenson, *What We're Fighting for Now*, 191.

9. For a philosophical discussion of this term, see Shue, *Climate Justice*. For a good set of resources and principles focused on international climate justice, see the work of the Mary Robinson Foundation–Climate Justice at www.mrfcj.org/.

10. Nash, *Loving Nature*, 218.

11. Cone, "Whose Earth Is It, Anyway?" 31. George Zachariah makes a similar point: "The poor in their collectivity is an epistemic community that creates oppositional knowledge. It is the seeing from the vantage point of the collectivity of the subalterns that has the potential to create oppositional knowledge." Zachariah, *Alternatives Unincorporated*, 101.

12. Rasmussen, "Environmental Racism and Environmental Justice," 19.

13. The roots of Christian ecological ethics are in Christian social ethics, an academic discipline that emphasizes the importance of understanding moral life in the context of systems and institutions. Social ethics has a long history of working not only with Christian communities but also political leaders, other communities, and all people of goodwill, suggesting that to be human is to be called to take challenges like racism, sexism, inequality, and injustice seriously on personal and structural levels. Ecological ethics continues these claims and extends them outward to add the issue of environmental justice, the degradation of the nonhuman world, and the extinction of species. Christian ecological ethics emphasizes that these challenges intersect and interlink, and all must be taken seriously by anyone seeking to live well in the world.

14. Yordy, *Green Witness*, 42. I have learned an enormous amount from Yordy's work, and in some ways this book has much in common with hers, particularly in its focus on peace and nonviolence as important environmental ideals. However, I have a fundamentally different audience than Yordy. She writes to the church; I write to the movement for climate justice. I suspect she would view my treatment of Christianity as she views Larry Rasmussen's: "at bottom, sociological rather than theological: a moral community rather than the body of Christ" (p. 143). I would concede that I am not writing an ethics that only makes sense in light of the specifics of Christian eschatological teaching but would argue that this remains a vitally theological project.

15. McFague, *New Climate for Theology*, 85.

16. Jenkins, *Future of Ethics*, 4. Jenkins engages a broad range of "inherited traditions" in response to a long list of "unprecedented" problems. This book attempts a smaller project: to apply the inheritance of nonviolent activism to the problem of climate change.

17. Ibid., 17.

18. All biblical quotations in this book are taken from the New Revised Standard Version.

19. Sallie McFague makes a similar turn to exemplary Christians in response to climate change. She calls her models "saints" and includes John Woolman and Dorothy Day alongside Simone Weil. She writes: "By following the clues in the lives of exceptional people—those called saints—one begins to understand, internalize, and perhaps to act in new ways. The saints 'scream' at us, the hard of hearing, and become living parables of a crazy, revolutionary, countercultural response to the reality they see before them: the world as radically interrelated and interdependent (an insight that contemporary science is also telling us)." McFague, *Blessed Are the Consumers*, 36.

20. King, *"In a Single Garment of Destiny,"* 6.

22. Pawel, *Crusades of Cesar Chavez*.

22. Sitkoff, *King*.

# PART I

Climate Change
and Nonviolence

# 1

## The Wicked Problem
## of Climate Change

In December 2015, 196 nations reached an agreement on climate change in Paris. They committed to limit global temperature increase "well below 2°C above pre-industrial levels and to pursue efforts to limit the temperature increase to 1.5°C."[1] Political leaders and analysts viewed the agreement as a historic success, with US president Barack Obama calling it "the best chance we have to save the one planet that we've got." Former vice president and longtime climate activist Al Gore was similarly excited: "The transformation of our global economy from one fueled by dirty energy to one fueled by sustainable economic growth is now firmly and inevitably under way."[2]

Others viewed the Paris Agreement as a failure that fell pitifully short of what is really required to responsibly address climate change. The journalist and activist Naomi Klein called the agreement "scientifically inadequate" because it lacks binding requirements to actually meet its goals. James Hansen, one of the most prominent climate scientists in the world, whose research informed the targets of 2°C and 1.5°C, called the agreement "just worthless words. There is no action, just promises."[3]

Both sets of commentaries are correct. From a political perspective, the Paris Agreement was a dramatic success because it secured widespread, public, and international commitment. From scientific and practical perspectives, however, the agreement is inadequate because it lays out no clear path to meet its own goals; and these goals, even if reached, would leave many of the world's people and species victimized by the rising seas and unpredictable storms of a changed climate. Paris was both a success and a failure, and which way one sees it depends on what kind of problem one thinks climate change is.

This same ambiguity is true of any conversation about climate change; there is always a great deal up for interpretation. Some facts are clear and incontrovertible—average global temperatures are higher than they used to be, there is less ice and snow at the poles, the sea is rising, and the ocean is increasingly acidic. These are real trends, and their rates are increasing. But the meaning of these facts and what should be done about them can be understood variously through natural science, environmentalism, activism for human justice, economics, politics, and religious traditions. As reactions to the Paris Agreement demonstrate, the same event can be an exciting development for climate politics while it is a disturbing sign for climate justice.

This chapter explores six common perspectives on climate change, assuming that each is valid and has something to offer to an understanding of this challenge. I then suggest a seventh approach, arguing that the privileged citizens of the industrialized world should also learn to see climate change as a case of structural violence. This seventh perspective, which frames the remainder of the book, suggests the potential of calling upon nonviolence to develop a thoughtful, long-term, and faithful response to this problem.

## SIX WAYS TO UNDERSTAND CLIMATE CHANGE

Everyone responding to the Paris Agreement in 2015 agreed on one thing: It is not a solution to climate change. Current international policies are, at best, a start along a path in the right direction, and no one pretends that they will reverse the rise of the oceans or undo the damage already done to the atmosphere. Climate change is, in fact, not the kind of problem that can have a simple solution. It is, instead, a "wicked" problem.

In 1973 Horst Rittel and Melvin Webber, scholars of urban planning, distinguished between two kinds of problems. "Tame" problems "are definable and may have solutions that are findable." By contrast, "wicked" problems "are ill-defined" and can never be solved. "At best, they are only re-solved—over and over again."[4] Rittel and Weber emphasized that urban planning in the late twentieth century was full of wicked problems. When one is negotiating space and relationships for diverse communities that struggle with unrest and inequity and must adapt to changing values and institutions, one does not finally solve any problem. The challenges are ambiguous, and there are never clear or final solutions.

Climate change is a similarly wicked problem. There is no solution; the climate will be changing because of human activity for many centuries to come. Human beings will be wrestling with and disagreeing about how to respond to this issue for the foreseeable future. That is the nature of a wicked problem.[5]

Wicked problems are also characteristically difficult to define. This aspect of climate change has become clear during the last two decades as scientists, activists, politicians, and academics have developed diverse perspectives and proposals. There is no one definition. Climate change is a scientific problem, an environmental problem, a human problem, an economic problem, a political problem, and a religious problem. Each perspective is important, and none is sufficient by itself. And yet they provide very distinct views of what is happening to the Earth's atmosphere, what the repercussions will be, and what should be done about it.

## A Scientific Problem

From one perspective, climate change is a physical phenomenon that has an impact on the Earth's atmosphere, oceans, and ecosystems. As such, it should be understood scientifically. Scientific measurement is essential to quantify the human impact on the climate. This includes, for example, 9.7 billion metric tons of carbon emitted in 2012, most of it from burning fossil fuels. Careful measurement of contemporary and historical air samples has shown that these emissions change the composition of the atmosphere, which contained approximately 280 parts per million of carbon dioxide ($CO_2$) in 1750 and 400 parts per million in 2015.[6] Thus, science helps us to understand that we change the climate when we burn fossil fuels to produce electricity, to transport ourselves around the world, and to manufacture the products that make our lives possible.

Scientific analysis also shows the industrialized food system as another driver of climate change. The Intergovernmental Panel on Climate Change (IPCC) estimates that agriculture is responsible for one-quarter of anthropogenic greenhouse gas emissions.[7] Modern farming depends upon fossil fuels to make fertilizer, process food, and ship it long distances. Forest land cleared to grow food is another part of the problem, as trees essential to the carbon cycle are cut and burned down so that crops can be grown instead. Industrial meat production has vastly increased the number of cows in the world, and cows release enormous amounts of methane, another climate-changing gas. These are problems that require rigorous scientific analysis to understand.

Scientific models are also essential to determine the atmospheric effects of the gases released by human food, energy, and transportation systems. Chemists explain that these gases all trap heat inside the Earth's atmosphere, creating a "greenhouse effect" that raises the quantity of the sun's radiation retained and reduces the amount reflected back into space. Atmospheric scientists measure temperatures and weather patterns and model future trends to determine how the climate will continue to change. According to the IPCC, which aggregates thousands of scientific studies, there has already been an increase in average global temperatures of at least 0.6°C, and, if current consumption trends continue, this could climb as high as 4.8°C by 2100. Alongside such an increase would be more extremes, with longer and more intense heat waves, and more rain in places that already tend to be wet and less in places that tend to be dry. The ocean will become more acidic while also continuing to rise up to 0.98 meter. Glaciers and ice packs will continue to shrink.[8]

Chemistry explains that climate-changing gases do not leave the atmosphere quickly. Atmospheric methane takes more than ten years to degrade. Nitrous oxide takes more than a century. Some of the $CO_2$ released today will remain in the atmosphere for hundreds of thousands of years. Thus, contemporary emissions will be changing the climate far into the future. Further emissions of these gases will increase the rate of such change.

We depend upon scientists to quantify and to predict climatic changes. For some, this leads to a hope that scientific research and engineering can also help us begin to solve, or at least address, this problem. For example, the Paris Agreement's goal of limiting temperature increases to 2°C came from scientific research, which also suggests that this limit would only be possible if atmospheric concentrations of $CO_2$ are lowered to 350 parts per million and kept there. The original source for this number is a scientific article by James Hansen and his colleagues, who used paleoclimatological analysis and models to argue that anything above 350 parts per million would be "too high to maintain the climate to which humanity, wildlife, and the rest of the biosphere are adapted." With characteristic caution, these scientists note that this target should be adjusted "as scientific understanding and empirical evidence of climate effects accumulate."[9] This is a scientifically determined goal, which continues to change as science develops and conditions change.

Some activists place even more faith in science by hoping for a technological path to reverse climate change. This could mean research to expand existing technologies, with the increased efficiency of solar, wind, and perhaps nuclear power fueling electricity and transportation infrastructures that could maintain the current standard of living in the developed world without fossil fuels. Other scientific solutions are even more revolutionary, such as proposals for "climate engineering," which would seek to recalibrate the atmosphere and the Earth's temperature by altering the amount of sunlight absorbed by the Earth or capturing $CO_2$ before it reaches the atmosphere. Most such proposals involve large-scale, global technologies. This requires trust that scientific experts can intentionally manage the atmosphere as a counterbalance to the careless and unintentional changes that have been made so far.

However, no single perspective is sufficient for understanding or responding to the wicked problem of climate change. No one suggests that science and technology can solve the problem of climate change alone, because the questions raised by it are not solely scientific. Any indictment of the industrial world's lifestyle has moral, socioeconomic, and cultural implications, as also does any decision about transitioning to new energy sources or deliberately changing the atmosphere.[10]

## An Environmental Problem

Although technological proposals about climate change are increasingly popular, others argue that the problem is most basically about consumption rather than engineering, and so call for fundamental moral and behavioral change. This has been the dominant approach of environmental activists, who see the changing climate as a sign of the faulty relationship between human beings and the rest of the natural world. In so doing, they present it as an environmental problem, connecting greenhouse gas emissions to other issues of pollution, overconsumption, and the endangerment of other species.

The first popular book about climate change, Bill McKibben's *The End of Nature*, put the issue in such terms. McKibben noted the dire effects that climate change poses for human communities, but he began with a more impressionistic reflection on the ways human beings are reshaping the atmosphere, noting that "the air around us, even where it is clean, and smells like spring, and is filled with birds, is *different*, significantly changed."[11] He carefully summarized scientific analyses of atmospheric conditions, but his focus was primarily on lamenting and decrying human carelessness. Understood environmentally, climate change is a problem because human beings are making the rest of the world less healthy, less beautiful, and less natural.

The iconic argument along these lines has been about polar bears, the charismatic megafauna most famously endangered by a warming planet. Visual representations of climate change by environmental organizations like the Sierra Club, the Nature Conservancy, and the World Wildlife Federation frequently focus on the destruction of the polar habitat, depicting an iconic white bear on a tiny ice floe in the middle of the ocean. These evocative pictures are based upon scientific research about the damage of climate change; polar bears hunt on Arctic Sea ice, which is appearing later in the fall and disappearing earlier in the summer each year. Their hunting season and their food supplies are both shrinking. As a specialized species that breeds slowly, polar bears are not well equipped to adapt to change, and so they face the very real threat of extinction in a warming world.[12]

Climate change is not only destroying the habitat of endangered species but also expanding the habitats of invasive species. Consider kudzu, a climbing vine that has grown to dominate much of the landscape of the southern United States since it was imported from Japan in the late nineteenth century. Kudzu thrives in edge habitats, boundaries between two different ecosystems. The vine is particularly prominent along roadsides, where it strangles trees and crowds out all other vegetation. The expansion of kudzu is currently only limited by the weather; it cannot survive cold winters and so does not grow in the northern or western United States. However, projections now suggest that as the climate warms, the range of this invasive species will increase. What is more, a 2014 study suggests that kudzu releases carbon from soils as it grows, and so increases atmospheric levels of greenhouse gases.[13] Climate change thus creates a positive feedback loop, increasing the range of an invasive species that, in turn, increases the rate of climate change.

Environmental discourse assumes that the extinction of a speices is a tragic loss and invasive species are a sign of degradation. The most common response is to call on people to change their attitudes and behaviors. Environmental organizations educate the public about the plight of the polar bear and the dangers of warming ecosystems in order to teach people that the problem is important, worthy of political action and personal sacrifice. When convinced, people are asked to eat less or no meat, to drive less or not at all, and to vote for policies that will slow or prevent the consumption of fossil fuels. Whereas those advocating technological solutions tend to focus on the ways climate change can be solved with ingenuity, environmentalists are more likely to stress that people must learn to care about the problem and the ecosystems it threatens.[14]

## A Human Problem

Many climate activists have recently distinguished themselves from the environmental movement, emphasizing the ways climate change is a threat to human communities rather than to other species and ecosystems. This, too, is based upon scientific research. According to the IPCC, climate change is likely to increase violent conflicts over resources, to reduce food supplies, to increase the uncertainty created by extreme weather events, to change the vectors of diseases and pests that threaten human beings, and to disrupt centuries-old agricultural traditions.[15]

The journalist David Roberts coined the term "climate hawk" in 2010 to name those who work to slow greenhouse gas emissions on behalf of humanity. This term incorporates those who "understand climate change and support clean energy but do not share the rest of the ideological and sociocultural commitments that define environmentalism as historically understood in the US." Four years later, a group in California formed a political action committee called Climate Hawks Vote, which is committed to electing candidates from any political party who prioritize this issue, emphasizing that "climate change is the greatest threat facing the next few generations of humanity, not just another Democratic issue."[16]

Climate hawks also call attention to the unjust distribution of the damage caused by climate change. Of course, all human lives are changed and threatened by a changing climate, but the harm is not evenly or justly shared. Instead, climate change is most harmful to those who are already the poorest and most marginalized, those who have emitted the fewest greenhouse gases because they tend not to travel broadly, use extensive amounts of power, or consume large quantities of meat.

The political analyst Richard Matthew demonstrates the injustice of climate change with the examples of Bangladesh and Sudan. Both countries are poor by global standards, with average annual incomes below $3,000 per person. Both have developed industrial activity only recently and in relatively small ways. Thus, they have contributed few of the greenhouse gases that are changing the global atmosphere. However, these two countries are nevertheless disproportionately suffering the consequences of human emissions. A low-lying coastal nation in South Asia, Bangladesh has 20 million citizens who live within 1 meter of elevation from current high tide levels. This means that rising sea levels, increasingly strong hurricanes, and flooding pose profound threats to Bangladeshis. By contrast, the North African nation of Sudan, which has been ravaged by civil war, is facing increasingly dry conditions, and growing deserts increase the risk of famine. Such resource scarcity then increases the chances of further violent conflicts. Mathews summarizes the trends of these two cases: "The costs of change are displaced onto the poor and weak, and the benefits of change are seized, often violently, by the rich and the powerful—whose unsustainable practices and values usually provided the rationale for change in the first place."[17] Climate change is a problem of justice because those who contributed the most to it are not those who suffer the most from it.[18]

As a wicked problem, climate change poses both environmental and human challenges; these are undeniably interconnected. However, the difference in emphasis is important. Traditional environmental rhetoric about climate change stresses that it harms ecosystems and creatures, having an impact on the entire world. Climate hawks instead stress that atmospheric changes hurt people and exacerbate existing injustices. Presenting climate change as an environmental problem calls primarily for increased concern about the interconnected systems of the Earth, while presenting climate change as a human problem calls primarily for outrage at the unfair risks facing human beings.

## A Political Problem

Most scientists, environmentalists, and climate hawks assume that the only way to address the immediate and dire threat of climate change is to change political structures and institutions. For example, the organization 350.org takes its name from scientific data but focuses its work on political change.[19] It organizes rallies around international climate summits, encourages citizens to lobby their national leaders to legislate limits on carbon emissions, and energizes local communities to divest from fossil fuels. This activist approach characterizes climate change as a challenge of democracy, leadership, and governance—in other words, a political problem.

Many activists assume that because climate change is a global problem, it requires a global political solution. In this view, the peoples of the world should unite around a joint effort to restrict the extraction and burning of fossil fuels and to create a new, more sustainable future. The most prominent step along these lines at the time of this writing has been the 2015 Paris Agreement, which created a common set of standards to measure political efforts aimed at reducing emissions. The fact that every country signed on creates a sense of global political community, putting pressure on every leader to do as much as possible. However, part of the success of this agreement came from the fact that it allows each nation to set its own goals. This made it possible for previously reluctant countries like the United States to sign but could also suggest that the true enforcement and motivation to politically address climate change must bubble up from somewhere lower than the entire global community.

As this book goes to press at the end of 2016, the Paris Agreement remains a key symbol of progress for many in the climate movement. However, its future is uncertain as the president-elect of the United States has suggested he does not believe that climate change is a problem and has discussed pulling out of this and many other international agreements.

The US has considerable international clout and is a considerable source of climate-changing gasses, so it remains to be seen whether other nations' commitments to Paris would hold without it. For some advocates of climate justice, this is a reason to focus even more attention on the international scale, advocating for renewed attention to climate change as a global problem and seeking to inspire leadership in other nations. Others in the US have a renewed domestic focus, seeking to use whatever tools are at their disposal to protect existing national policies and continue fighting climate change with or despite a new presidential administration.

Former US vice president Al Gore tends to treat climate change as a national issue. He changed the conversation about the issue with his film and book, both titled *An Inconvenient Truth*, arguing that the United States must lead the world's action on climate change. Gore patriotically insists that the country that created a democratic Constitution and landed on the moon has the best chance of decisively addressing the twenty-first century's greatest challenge. Near the end of his book, he writes, "Now it is up to us to use our democracy and our God-given ability to reason with one another about our future and make moral choices to change policies and behaviors."[20] This is a call to national political action, and Gore has since advocated congressional legislation that would cap greenhouse gas emissions and increase support and subsidies for renewable energy sources. Of course, Gore also celebrated the 2015 Paris Agreement and advocated ratification of its predecessor, the Kyoto Accord, but his primary interest is inspiring US citizens to lead the way with a national response to climate change.

Still, others emphasize that climate change is a local political issue. In 2005 Seattle mayor Greg Nickels asked other mayors to commit to lowering their emissions by changing municipal land-use policies, informing their citizens, restoring urban forests, and limiting greenhouse gas emissions. By 2009 a thousand mayors had signed on, and Nickels argued that "a successful plan in this country for reducing our energy consumption begins in cities and local communities. We are leading by example in the fight against global warming and representing America to the world."[21] Seattle continues to work at leading by example, and in 2013 it adopted a Climate Action Plan that predicted some of the Paris goals by committing the city to zero net greenhouse gas emissions by 2050.[22]

Although they work on different scales, these efforts share a common assumption that climate change is a political problem, and so they organize citizens and leaders to take action—locally, nationally, or globally. A political problem calls for laws and treaties that discourage greenhouse gas emissions and promote alternative energy and conservation.

## An Economic Problem

Political proposals have their critics, and the most vocal of these tend to instead advocate a market-based, economic approach. The British philosopher Roger Scruton worries that most political responses to climate identify "unreal targets, pursued in ignorance of the means to achieve them, and without any conception of how the attempt to do so will impinge on popular sentiment, on competing goals and on the many other factors that wise government must consider." Scruton worries that political proposals are "too often made without being priced," and he insists that any attempt to control and limit human behavior without a clear analysis of costs and benefits is doomed to failure.[23] Thus, he argues, climate change should be understood as an economic rather than political problem.

Fred Smith, founder of the Competitive Enterprise Institute think tank, takes a similar approach, assuming that policies restricting fossil fuels and subsidizing other energy sources will limit innovation and increase human suffering by slowing economic growth. So, he argues, "We should remove regulatory barriers that limit innovation and technological advance. . . . The policies that are best for the ecology of the earth are those that are best for the economy of the earth. No policy that harms people can help our planet."[24]

Others agree that climate change is an economic problem but seek to solve it with drastic changes that require more rather than less political regulation. The activist Naomi Klein argues that capitalists like Scruton and Smith are right to see climate politics as an attack against existing economies, but she insists that this attack is right and necessary. She calls progressive climate activists to embrace the economic argument "that the real solutions to the climate crisis are also our best hope of building a much more enlightened economic system—one that closes deep inequalities, strengthens and transforms the public sphere, generates plentiful, dignified work and radically reins in corporate power." For Klein, climate change is an economic problem caused by "the reckless form of 'free trade' and 'the growth imperative' that define contemporary capitalist economies." The solution is a different kind of economics—with higher taxes on polluters, higher charges for those who extract fossil fuels, and huge investments in cleaner energy.[25]

Another progressive economic argument came from the heads of state of Bolivia, Cuba, Dominica, Honduras, Nicaragua, and Venezuela in 2009, when they asserted that "capitalism is leading humanity and the planet to extinction," with climate change as a key example. They emphasized the unjust distribution of climate change's effects on those who have done the least to cause it and insisted that this creates an economic responsibility: "Developed countries should pay off their debt to humankind and the planet; they should provide significant resources to a fund so that developing countries can embark upon a growth model which does not repeat serious impacts of the capitalist industrialization."[26] This assertion of "climate debt" is prominent in Latin America and Africa, but has yet to be validated by leaders in the developed nations of Europe or North America.

Whether calling for less regulation or a stronger governmental role in distributing the risks of climate change, these arguments all share a view of the problem as fundamentally economic.

## A Religious Problem

For communities of faith, climate change is not just scientific, environmental, human, political, and economic; it is also a religious problem. The Western monotheistic traditions of Judaism, Christianity, and Islam share strong doctrines of creation, asserting that God made the world and declared it good. In light of such teaching, the careless and unintentional altering of climatic balance is disrespectful at best and blasphemous at worst. The Eastern traditions of Buddhism, Hinduism, and Taoism are less likely to emphasize creation, but they share a sense that everything is interconnected and any harm done to any part of the world has negative consequences for all. From this perspective, the degradation of ecosystems and communities is not only irresponsible but also self-destructive.

In January 2014, a group of Buddhist teachers from all over the world released a statement on climate change titled "The Earth as Witness." They used the foundational four noble truths of Buddhism to frame the issue and its solution. The first noble truth, acknowledging the reality of suffering, offers a way to recognize the risks, injustices, and present harm caused by a changing climate. The second noble truth, that suffering arises from desire, leads to an argument that "craving, aversion, and delusion" are the root causes of the consumption that changes the climate. The third noble truth, that human beings can learn to overcome desire, offers hope that "we can create more equitable, compassionate, and mindful societies that generate greater individual and collective well-being while reducing climate change to manageable levels." Finally, the fourth noble truth presents a path away from destructive desire and consumption. The statement concludes that the work of restoring balance to the Earth honors "the great legacy of the Dharma and fulfill our heart's deepest wish to serve and protect life."[27]

A Hindu statement released during the 2009 Parliament of the World's Religions in Melbourne similarly links climate change to core ideas within a religious tradition. Hinduism has deep roots in sacrificial rituals that attune practitioners to the ultimate unity of the cosmos, and so it is no surprise that the statement emphasizes the need for sacrificial action: "As one-sixth of the human family, Hindus can have a tremendous impact. We can and should take the lead in Earth-friendly living, personal frugality, lower power consumption, alternative energy, sustainable food production and vegetarianism." Such sacrifices offer a path toward "a global consciousness that replaces the present fractured and fragmented consciousness of the human race."[28]

Islamic faith tends to emphasize sacred texts much more than Hinduism or Buddhism, and so it makes sense that Muslim responses to climate change frequently focus on the teachings of the Qur'an. For example, the British Muslim Fazlun Khalid draws upon the thirtieth *surah* of the Qur'an to articulate his understanding of the problem:

> Corruption has appeared in both land and sea
> Because of what people's own hands have brought about
> So that they may taste something of what they have done
> So that hopefully they will turn back.[29]

Inspired by this verse and by his faith more broadly, Khalid founded the Islamic Foundation for Ecology and Environmental Sciences and calls other Muslims to avoid corrupting the Earth and instead become the stewards that he believes their faith calls them to be.

Islam, Hinduism, and Buddhism are, respectively, the second-, third-, and fourth-most-populous religions in the world. The largest global religion is Christianity, and this book proposes a set of resources from this tradition. More relevant here is the broader point that every religious community offers responses to climate change. Every religious community adds another set of perspectives from which to understand this multifaceted, wicked problem facing twenty-first-century human beings.[30]

## Wicked Thinking

The point of presenting these diverse perspectives is not to force a choice between them but rather to see that climate change is the kind of problem that cannot be fully understood from one vantage point. Science is essential to grasp what is happening to the atmosphere, but it is incomplete without the moral arguments of environmentalists and social justice advocates. These arguments require careful economic and political analysis if they are to have practical effects. All these perspectives have much to learn from the ancient wisdom of the world's religious traditions, but these traditions must also be open to learning about this contemporary, evolving challenge. As a wicked problem, climate change can never be fully understood from one perspective.[31]

To wrestle with a wicked problem is to encounter hard questions that will never have clear answers. After the argument of this chapter and the next, the remaining chapters use the witnesses of nonviolent Christians to wrestle with a series of questions suggested above: How should wealthy people in the industrialized world respond to the fact that our actions are changing the climate (chapter 3)? How should concerned people prioritize between environmental advocacy for the Earth's ecosystems and the particular injustices afflicting marginalized human communities (chapter 4)? In a pluralistic society and world, how can religion inform and enhance a moral response to climate change and the questions it raises about global economics (chapter 5)? Should environmentalists and social justice advocates support proposals to intentionally engineer the climate on a large scale (chapter 6)? How should privileged persons in the industrialized world respond to the argument that we owe a "climate debt" to the developing world (chapter 7)?

None of these questions will be finally answered—as explained above, a wicked problem does not lend itself to such closure—but each is explored in the chapters that follow in an attempt to advance moral consideration, engaged debate, and meaningful witness. Before moving on to this analysis, however, it is necessary to first explore the seventh perspective on the wicked problem of climate change, which guides the rest of this book.

## A PROBLEM OF STRUCTURAL VIOLENCE

Climate change is not only a scientific, environmental, human, political, economic, and religious problem. It is also a problem of structural violence.

To understand this claim first requires an understanding of violence. A useful, if abstract, definition comes from the French philosopher Emmanuel Lévinas, who writes, "Violence is to be found in any action in which one acts as if one were alone to act: as if the rest of the universe were there only to *receive* the action; violence is consequently any action which we endure without at every point collaborating in it."[32] Thus, for Lévinas, violence occurs when anyone behaves as if they are the only one who matters—ignoring other people, other creatures, and the rest of the world. From the perspective of a perpetrator, violence is an act of selfishness, doing something to someone or something else without considering its effects upon them. From the perspective of the victims, violence is something done to them without their consent.

This is a useful perspective, but it is also quite broad—Lévinas himself points out that "nearly every causality is in this sense violent."[33] Therefore, I refine the definition slightly to say that violence is an act undertaken by a human being who behaves as if they were alone and therefore causes harm to another creature. This includes behaviors that one tends to think of as violent—hitting or shooting another person, shouting abuse—but also others that might not as intuitively fall into this category—dumping toxic pollution into a river, mistreating a nonhuman animal, neglecting to consider another person's feelings. Violence is selfish action that causes harm. According to this definition, climate change is violence.

Furthermore, the violence of climate change is distinctly structural. The idea of "structural violence" has been widely used to characterize the problems of patriarchy, racism, and classism. Introducing the term in a 1969 essay, the peace researcher Johan Galtung explains structural sexism and classism: "When one husband beats his wife, there is a clear case of personal violence; but when one million husbands keep one million wives in ignorance, there is structural violence. Correspondingly, in a society where life expectancy is twice as high in the upper as in the lower classes, violence is exercised even if there are no concrete actors one can point to directly attacking others."[34] This same dynamic exists in expressions of racism. A Ku Klux Klan member who burns a cross and a police officer who is prejudicially forceful toward an African American are committing acts of direct violence. But racism is also structural; it is discriminately harder for people of color in the United States to become educated and employed and to establish multigenerational wealth because of centuries of slavery, segregation, and prejudice.[35] In contrast to the direct violence of abuse, assault, or murder, structural violence is caused indirectly by social systems; it is no one person's responsibility and so no one person's fault. It has no single architect and no direct cause, but it is nevertheless violence—a selfish expression of power that harms others.

To identify climate change as a problem of structural violence, then, is to observe that some human beings are selfishly altering the atmosphere, and these changes are causing pain and suffering. This complements many of the perspectives named above; climate change is a human problem because it is about structural violence against people. With a broader awareness of suffering among all creatures, climate change can also be seen as a problem of environmental violence because it hurts other species and ecosystems. And yet the structural violence of climate change can only be understood scientifically because the types of harm it causes are often far distant from their causes. Furthermore, existing socioeconomic and political structures do not require those who emit $CO_2$ or eat industrially raised meat to pay for the expansion of kudzu or droughts in the Sudan; this could only be changed with new structures.

The Christian ethicist Cynthia Moe-Lobeda insightfully explores climate change as a form of structural violence in her book *Resisting Structural Evil*. Climate change, she writes, "degrades, dehumanizes, damages, and kills people by limiting or preventing their access to the necessities for life or for its flourishing."[36] She further notes an "insidious characteristic" of structural violence: "its tendency to remain invisible to those not suffering from it."[37] Wealthy people in the industrialized world can still deny the problem, and thus can still ignore the violence of climate change. The literary scholar Rob Nixon captures this reality when he identifies climate change as a form of "slow violence"—pain and harm that is "neither spectacular nor instantaneous, but rather incremental and accretive, its calamitous repercussions playing out across a range of temporal scales."[38] The habitats of polar bears shrink gradually; the waters on the Bangladeshi coast rise slowly. Those of us who are disproportionately causing climate change are far away enough from the worst consequences that we can ignore them, at least for now.

## Repenting of Climate Change

In response to the structural violence of climate change, Moe-Lobeda calls her readers to learn to "'see' the structural sin of which we are a part, in order that we might repent of it, renounce it, and resist it."[39] This move to repentance is crucial, as privileged people must learn to think and behave differently, to remake social systems, and to turn away from violence.

Repenting of the structural violence of climate change requires concerted moral action. Those complicit in the violence must first acknowledge their complicity. Despite the fact that no one intends to cause climate change, everyone is part of the problem. When I drive my car or drink milk produced on a factory farm, I am not thinking about climate change, but my actions lead to the release of $CO_2$ and methane into the atmosphere. When energy executives prioritize the security of jobs and the financial value of fossil fuels, they may intend only to help their employees and investors, but they are nevertheless degrading ecosystems and human lives. When politicians refuse to restrict emissions, they may be responding to the short-term needs of their constituents, but they are nevertheless causing long-term harm to the planet's ecosystems.

Admitting this complicity in climate change can be empowering. When those of us who contribute the most to climate change realize that we have particular influence over the structures that cause the problem, we also learn that we have the capacity to change them for the better. Moe-Lobeda emphasizes that such agency is vital; people must understand that we can change our lives and the structures in which we live. When

facing political and economic systems that seem to make continued climate change inevitable, people need to recognize that these systems "were
constructed by human beings and therefore can be changed by them.
The neoliberal global economy—including its manifestation in national
economies—was constructed by people. It can therefore be replaced."[40] If
those who benefit disproportionately from the forces changing the climate
acknowledge our power, we can begin to use it for the sake of other people
and the whole Earth community, we can begin to share our power and to
learn from those who have been excluded. We can stop acting as though
we were alone in the world and can begin cooperating with others.

Along these lines, the theologian Ernst Conradie proposes that everyone
concerned about climate change can learn from the South African response
to apartheid, in which the creation of a new society was made possible partly
because many of those guilty of deep structural racism confessed that guilt.
South Africa's Truth and Reconciliation Commission was by no means perfect, and the problem of structural racism is by no means solved in that
nation. However, new relationships and new kinds of relationships have
been formed because the guilty confessed and everyone shared a willingness to create new systems. Conradie hopes that, "like the beneficiaries of
apartheid, it may be possible, by God's grace alone, to accept and specify
one's responsibility towards causing climate change, to recognise one's guilt
and show remorse."[41] Such confession and remorse may open the path to
imagining a new set of structures that heal rather than degrade the planet.

## A New and Familiar Problem

This book is based on a hope that echoes Conradie's, a hope that the strategies people have used to repent of racism, sexism, classism, and other
problems of structural violence will also be useful in repenting of climate change. For this reason, I disagree with one aspect of Cynthia Moe-
Lobeda's account of climate change. She stresses the ways in which this is
a brand new and unprecedented problem, while I believe it is important
to focus on the continuity between climate change and other examples of
structural violence.

Moe-Lobeda writes that "climate change presents new theological problems for our young and dangerous species" and that the moral

challenge we face is one "the likes of which the world has never before known."[42] Never before, she insists, has the entire global system of life been so fundamentally changed and threatened by the consumption of some members of one species. Never before have human beings been so urgently required to rethink their economic, political, and social structures, and their attitudes toward one another and the rest of the world.

This claim of uniqueness is very common in writing about climate change. The Christian ethicist Willis Jenkins agrees that climate change is a new kind of ethical problem, involving "dimensions of human action without precedent in our traditions and institutions of justice."[43] The activist Naomi Klein's 2014 book about climate change was titled *This Changes Everything* precisely to signal that industrialized societies must change their thinking and structures in basic ways.[44] The journalist Wen Stephenson insists that the path to climate change requires the world to "wake up" to the crisis, "intellectually, morally, and spiritually, as the most fundamental and urgent threat humanity has ever faced."[45]

I agree that climate change calls for innovative economic, political, moral, and religious thinking and action. However, this is a wicked problem, and so there is never just one way to understand it. The structural violence of an altered atmosphere is certainly new; but it is also very old. Meaningful response to climate change will come more helpfully from emphasizing the familiarity of this problem. Concerned people know that we face new challenges, but in order to face them well, we need reminders that we have resources from the past with which to do so.

One argument for the familiarity of climate change could be based on scientific precision. The climate has been changing as long as there has been an atmosphere on Earth. Climatic change billions of years ago created the relatively even temperatures that made life possible. Life then began to change the climate, as microorganisms altered atmospheric chemistry. Plants evolved the process of photosynthesis, which plays a central role in the carbon cycle, sustains the ozone layer, and makes animal life possible. As evolution continued, there were warm periods and ice ages. Although one species has never before altered the atmosphere on the scale at which humanity is now doing so, a changing climate is not a new phenomenon.

Even anthropogenic climate change can be understood as familiar. As a scientific problem, climate change requires the same tools that human beings have been using for centuries to understand the seasons and weather. As an environmental problem, climate change requires the same attention to natural systems and interconnection required by the hole in the ozone layer, industrial pollution, and species extinctions. As a social justice problem, climate change is inseparable from and therefore requires similar moral work as the challenges of colonialism, globalization, racism, and economic justice. As a political problem, climate change challenges systems of global, national, and local governance just like other forms of structural violence. As an economic problem, climate change requires careful attention to how resources and risks are and could be distributed. As a religious problem, climate change asks people of faith to think carefully about how what they hold sacred changes how they live in and care for the world, a question that is older than the Bible's book of Genesis.

As a wicked problem of structural violence, climate change is not unprecedented. Racism, classism, sexism, ethnocentrism, and heterosexism are also wicked problems of structural violence, and human beings have been struggling with these problems for centuries—in at least some cases, for as long as humans have been a species. Climate injustice is a new addition to the list, but it fits into the preexisting category of structural violence. Furthermore, because the types of harm inflicted by climate change are primarily borne by the poor and marginalized while primarily being caused by the rich and powerful, climate change continues and exacerbates other forms of structural violence. As the ethicist Melanie Harris notes, "The crisis of climate change facing us all suggests that the logic of domination that has functioned to privilege white men and white communities over communities of color is not working."[46] In this sense, climate change is not new.

I emphasize the familiarity of climate change because it is empowering to know that those who came before us developed tools with which to respond to problems like climate change when they fought racism, sexism, heterosexism, and classism. Of course, no resistance to structural violence is an unqualified victory. Racism, sexism, heterosexism, and classism still exist and continue to take new shapes as the world changes. But each case offers stories of development and growth that make room for hope. Racism still exists in the United States in many horrific forms, but segregation and slavery are illegal. Sexism remains pervasive in the predominant patriarchal society, but the rights of women to vote, control their own bodies, and direct their own lives are far more guaranteed than they were a hundred years ago. Heterosexuality still dominates the norms of most human cultures, but the past decades have seen drastic changes in recognizing diverse expressions of sexuality and gender. Economic inequality is a deeply destructive form of structural violence, but the twentieth century saw the creation of a basic social safety net in the United States, with more health care and food security extended to the poorest citizens. None of these is a success story; wicked problems do not allow solutions. But each offers examples of accomplishment.

A movement for climate justice requires good news from the past in order to look with realistic hope into the future. Climate change is a wicked problem, but good people have wrestled with wicked problems before, and people today have much to learn from them. The next chapter turns to one such legacy: nonviolent resistance.

# NOTES

1. Conference of the Parties, "Paris Agreement."
2. Suzanne Goldenberg et al., "Paris Climate Deal: Nearly 200 Nations Sign in End of Fossil Fuel Era," *Guardian*, December 12, 2015.
3. "Paris Climate Deal Is Agreed—but Is It Really Good Enough?" *New Scientist*, December 12, 2015, www.newscientist.com/article/dn28663-paris-climate -deal-is-agreed-but-is-it-really-good-enough/; Anna Maria Termonti, "Naomi Klein Calls Paris Climate Agreement 'Scientifically Inadequate,'" *Current* (CBC Radio), December 15, 2015, www.cbc.ca/radio/thecurrent/the-current-for -december-15-2015-1.3365556/naomi-klein-calls-paris-climate-agreement -scientifically-inadequate-1.3365564.
4. Rittel and Webber, "General Theory of Planning," 160.
5. In a way that has been influential upon and is hopefully compatible with the approach of this book, Willis Jenkins argues that understanding climate change as a wicked problem justifies a pragmatic approach to contemporary Christian ecological ethics. As discussed below, my approach differs from his emphasis on creativity over tradition, but my work nevertheless owes a debt to his methodological proposals. See Jenkins, *Future of Ethics*, chap. 4.
6. "Global Carbon Emissions," http://co2now.org/.
7. IPCC, "Summary for Policymakers," 2014, 25.
8. IPCC, "Summary for Policymakers," 2013, 19–29.
9. Hansen et al., "Target Atmospheric $CO_2$," 228–29.
10. See Robock, "Geoengineering May Be Bad"; British Royal Society, *Geoengineering the Climate*; and Clingerman and O'Brien, "Playing God."

11. McKibben, *End of Nature*, 18, 72.

12. Scientific accounts of polar bears upon which such environmentalists' claims are based include those by Stirling and Derocher, "Effects of Climate Warming"; and Derocher et al., "Rapid Ecosystem Change."

13. "Clemson Scientist: Kudzu Can Release Soil Carbon, Accelerate Global Warming," 2014, http://newsstand.clemson.edu/mediarelations/clemson-scientists-kudzu-can-release-soil-carbon-accelerate-global-warming/; Bradley, Wilcove, and Oppenheimer, "Risk of Plant Invasion."

14. For an account and analysis of the ways environmentalism emphasizes care about ecosystems, see Blanchard and O'Brien, *Introduction to Christian Environmentalism*, chap. 2.

15. IPCC, "Summary for Policymakers," 2014, 4–8.

16. David Roberts, "Introducing 'Climate Hawks,'" 2010, http://grist.org/article/2010-10-20-introducing-climate-hawks/; "Climate Hawks Vote," www.climatehawksvote.com.

17. Matthew, "Climate Change and Security," 273.

18. For more accounts of the impact climate change has upon cultures and nations that have relatively little responsibility for it, see especially Parenti, *Tropic of Chaos*.

19. The organization explains its name on a Web page titled "The Science," which cites James Hanson as an authority; see http://350.org/about/science/.

20. Gore, *Inconvenient Truth*, 296. Writing about Gore and other climate activists, Eric Pooley notes that although rhetoric frequently balances personal as well as political actions in this way, "fighting climate change in an industrial society requires political action at the local and—especially—the national level." Pooley, *Climate War*, x.

21. US Conference of Mayors, "1,000th Mayor—Mesa, AZ, Mayor Scott Smith—Signs the US Conference of Mayors' Climate Protection Agreement," October 2, 2009, www.usmayors.org/pressreleases/uploads/1000signatory.pdf.

22. "Seattle Climate Action Plan," 2013, www.seattle.gov/environment/climate-change/climate-action-plan. For an extensive ethical analysis of global, national, and local political efforts, see Martin-Schramm, *Climate Justice*.

23. Scruton, *Think Seriously About the Planet*, 58.

24. Smith, "Risk–Risk Approach," 251.

25. Naomi Klein, "Capitalism vs. the Climate," *The Nation*, November 9, 2011, www.thenation.com/article/164497/capitalism-vs-climate#. This argument is further developed by Klein, *This Changes Everything*.

26. "The Declaration of Cumaná," April 29, 2009, http://nacla.org/news/declaration-cumaná,.

27. "The Earth as Witness: International Dharma Teachers' Statement on Climate Change," January 8, 2014, www.oneearthsangha.org/articles/dharma-teachers-statement-on-climate-change/.

28. "Hindu Declaration on Climate Change."

29. Fazlun Khalid, "Global Warming: An Islamic Perspective," 2007, www.ifees.org
.uk/download.php?id=36.

30. For a helpful set of ecumenical perspectives on climate justice and accounts of
diverse religious communities acting on the issue, see Kim, *Making Peace with
the Earth*.

31. Mike Hulme expresses this argument well, suggesting that framing climate
change as "a mega-problem in need of a mega-solution" creates "a log-jam of
gigantic proportions." He argues that we should instead learn to see climate
change as a malleable idea that can be "used to rethink and renegotiate our
wider social goals and how and why we live on this planet." This is one way of
understanding the task of this chapter and this book: understanding climate
change as violence in order to imagine and nurture a more nonviolent world.
Hulme, *Why We Disagree*, 333, 362.

32. Lévinas, *Difficult Freedom*, 6. I learned this definition from Robert Brimlow and
recommend his discussion of it in *What About Hitler?* 133–36.

33. Lévinas, *Difficult Freedom*, 6.

34. Galtung, "Violence and Peace Research," 171.

35. See especially Townes, *Womanist Ethics and Evil*.

36. Moe-Lobeda, *Resisting Structural Evil*, 72.

37. Ibid., 61.

38. Nixon, *Slow Violence and Environmentalism*, 2.

39. Moe-Lobeda, *Resisting Structural Evil*, 61.

40. Ibid., 230.

41. Conradie, *Church and Climate Change*, 88. Moe-Lobeda similarly hopes that
churches will make genuine repentance possible: "Could it be that worship
that empowers the people of God for social and ecological healing will include
profound lament for the ways in which our lives unwittingly endanger Earth's
life-systems and vulnerable neighbors far and near?" Moe-Lobeda, *Resisting
Structural Evil*, 262.

42. Ibid., 54, 299.

43. Jenkins, *Future of Ethics*, 1.

44. Klein, *This Changes Everything*.

45. Stephenson, *What We're Fighting for Now*, ix. Interestingly, Stephenson also calls
the climate justice movement to model itself on the abolitionist movement; see
pp. 30ff.

46. Harris, "Ecowomanism," 7.

# 2

## Nonviolent Resistance

Nonviolence is countercultural, because war and violence are part of how most human cultures define themselves. Consider the typical history of the United States, which moves from the Revolutionary War, to the Civil War, to world wars, and to terrorism—all defining the nation by its uses of violence to free itself and affirm democracy. Consider the quintessential fictional heroes created by popular media, like Luke Skywalker, Katniss Everdeen, and Harry Potter—good people who become experts at violence in order to stand up against evil people. Violence is part of culture, and so opposing violence is countercultural. To imagine living and defining oneself without violence is to go against the norm.

In 1906 the philosopher William James gave a speech on this subject titled "The Moral Equivalent of War." It was an optimistic time; in the years before World War I, many intellectuals believed that the international community was on a path to make war illegal and unacceptable. James was trying to anticipate the problems of this peaceful world, and he considered the fact that meaning and discipline tend to come from war. Violence has become "a sort of sacrament" and "a permanent human *obligation*." Centuries and millennia of war have made it seem normal, and "the militarily patriotic and romantic minded" have learned to define themselves, their communities, and their nations with stories about war and violence. So, James suggested, those who oppose war must create a "substitute for war's disciplinary function"—a "moral equivalent of war."[1] Cultures must find some other struggle with which to define themselves.

James's answer to this problem is a bit shocking to twenty-first-century ears because he proposed that all young people should be drafted into an "army enlisted against *Nature*." He imagined young adults being taught to mine coal or catch fish or build skyscrapers in order "to get the childishness knocked out of them, and to come back into society with healthier sympathies and soberer ideas," after having "done their own part in the immemorial human warfare against nature" (emphasis in the original).[2] The substitute for war against other people, for James, was a "war" against the nonhuman world.

The evidence given in the last chapter suggests that such a "war" is already being waged. Glaciers and polar bears and kudzu all reveal that humanity has gotten very good at violence against the natural world. But climate refugees and flooded coastal communities suggest that violence against nature is also, inevitably, violence against human beings. Climate change teaches a lesson that much of the world had not learned in 1906; there is no way for human beings to "wage war" on the natural world without also harming themselves. There is no peace for people who remain violent against nature.

Mohandas Gandhi found a different moral equivalent to war in nonviolence—in the commitment to resist violence without using violence. The same year William James gave his speech, Gandhi opposed a law that required Indians in South Africa to register with the government. He publicly refused to register and urged other Indians to do the same. During the next seven years, thousands were beaten and jailed for their disobedience. The resisters openly faced these punishments, and the fact that the state was violent while the protesters were peaceful drew a public outcry. South African leaders eventually negotiated a compromise. Gandhi then took his strategy of nonviolent resistance home to India and used it to oppose British colonization for the next forty years. Eventually, the Indian people won independence without fighting a war against the British.[3]

Like war, nonviolence requires discipline; those who protested with Gandhi had to be rigorously trained so that they would not fight back, even when the police beat them. Like war, nonviolence creates a sense of community; the nation of India to this day prides itself on an origin that did not require the violent overthrow of its oppressors. Like war, nonviolence creates a legacy of heroism; Gandhi was named Mahatma, or "Great Soul," during his lifetime, and he continues to be revered in India and around the world.

William James was right that opposing violence requires significant cultural change.[4] But opposing violence does not require a "war" against nature—or anything else. Instead, this chapter argues, the best way to resist violence is to side with human beings and with nature, to learn the counterculture of nonviolence from Gandhi's movement and the movements it has inspired.

This chapter's argument narrows as it develops—from an overarching view of nonviolence, to Christian expressions of nonviolence, and then to the particular work of five witnesses from the United States. Each of these narrowings is meant to clarify and focus, but none is meant to exclude. I am not arguing that nonviolence is the only acceptable response to the violence of the world; that is a broad claim beyond the scope of my work. This chapter does not even argue that all responses to climate change must be nonviolent; the problem is too wicked to lend itself to such simplicity. Nor do I believe that all nonviolence should be motivated by Christian faith; people of many traditions and no tradition have contributed enormously to the nonviolent tradition.[5] Finally, the argument is not that all Christians must embrace nonviolence as it has been modeled in the United States; there are many other long-considered and insightful Christian responses to violence. The argument is, instead, simply this: All who are concerned about climate change have something to learn from this Christian tradition of nonviolence.[6]

## DEFINING NONVIOLENCE

The previous chapter defined violence as selfish action that causes harm. I act violently when I act as if I am the only one who matters, hurting others through carelessness or maliciousness. Nonviolence is the opposite because it insists upon care for and attention to others. This means resisting violence while refusing to use violence.[7]

### Resisting Violence

Nonviolence is often assumed to mean passivity, a refusal to act. The political scientist Gene Sharp, who extensively studied Gandhi's movement and many other similar ones, insists that this is not true. Nonviolence, he writes, "is *action* that is nonviolent," and thus it consists of active resistance.[8] Furthermore, nonviolence seeks to coerce those who perpetrate and allow violence to change what they are doing. In his classic *The Methods of Nonviolent Action*, Sharp names 198 distinct methods for nonviolent action, from picketing to refusing to pay taxes to sit-ins.[9]

Gandhi famously resisted any use of violence, but he less famously said that he would rather have violence than inaction. In 1924 he wrote, "Between violence and cowardly flight, I can only prefer violence to cowardice. I can no more preach nonviolence to a coward than I can tempt a blind man to enjoy healthy scenes."[10] He taught Indians to nonviolently resist British colonialism; but to do so, he first had to convince them to recognize the violence of colonization and oppose it, to be willing to fight against injustice. The first step, he insisted, is to resist violence.

On a personal level, resisting violence means standing up for oneself when one is attacked or hurt and standing up for those around us who are being attacked or hurt. People who walk away from an abusive relationship or defuse an escalating argument are resisting violence. On a political level, resisting violence means standing in the way when a government acts violently. People who march in opposition to an unjust war or call upon their elected representatives to invest more in diplomacy rather than weapons of mass destruction are resisting violence. On a global level, resisting violence means responding when one hears about injustice, oppression, or structural violence anywhere in the world. People who boycott a product or divest their resources from a company because of harm caused to others are resisting violence.

When violence is structural, resisting violence is about creating structural change, opposing the systems that leave some people destitute, disenfranchised, or dejected. Nonviolent resistance should lead to better structures, to a more equitable, more democratic, and more empowering society. As Sharp and his colleague Joshua Paulson write, "Real and lasting liberation requires significant changes in the power relationships within the society. . . . Liberation should mean that the members of the previously dominated and weak population obtain greater control over their lives and greater capacity to influence events."[11] Resisting violence means changing the systems that create violent outcomes.

In the twenty-first century, nonviolence must include resisting the violence of climate change. People should resist the temptation to use fossil fuel energy or consume industrial foods that they do not need. Citizens should march and protest when governments fail to pass laws limiting overproduction and overconsumption. All human beings should find ways to help those who have been driven from their homes and livelihoods by the changing atmosphere. To resist violence is to seek climate justice.

## Refusing to Use Violence

There are many ways to resist violence. If I feared a violent attack on my home or my family, I would call my city's police, who are armed so that they can prevent violence with the use or threat of violence. When the United States feels threatened by an opposing army or terrorist group, it increases its investment in the military and homeland security, which are highly prepared to use violence in order to oppose any threat against the nation. What makes nonviolence unique is that it does not appeal to violence—it refuses to resist violence with violence. To strike out violently runs the risk of killing the opponent, and even nonlethal force reduces the possibility of future dialogue. So, nonviolence stands up against threats and injustice without resort to guns, fists, or the threat of harm to the opponent. This form of resistance seeks to win without forcing the other side to lose in a final or ultimate way.

For many nonviolent activists, the refusal to use violence is motivated by a refusal to believe that the world is simplistically divided into good and evil people—us and them. Instead, nonviolence is based on a belief that no one is perfect and everyone can become better. To know that no one is perfect is to know that I myself have limitations and make mistakes, and this makes me cautious about using violence. Because all people make mistakes, I should be wary of allowing myself (or, indeed, my nation) easy access to weapons that could cause irreparable harm. I can never be sure that such tools would be used wisely or well.[12] At the same time, to know that everyone can be better changes the way I treat my opponents. If I understand that those with whom I disagree, though flawed, can become better, then I will be cautious about ending their lives or putting them on the defensive with a threat of harm.

Another reason many activists refuse to use violence is because they worry about its long-term effects. Nonviolence is the sensible choice when one cannot win by violence or when one worries that violence would result in too much destruction to one's home or one's relationships. In South Africa Gandhi organized a small minority of Indians who would otherwise have been crushed by the government, so nonviolence was strategically wise. In India there were enough native Indians to overthrow the minority of British colonizers, but Gandhi worried about the future of a state founded on violence, and so he found another path. In both cases he believed that nonviolence would lead to a better future than violence.

Therefore, nonviolence avoids harming the other out of awareness of one's own limits, respect for the humanity of the opponent, and a strategic desire for the most positive outcome. The Christian activist and theologian Ronald Sider accurately expresses all these justifications for nonviolence, noting that nonviolence "respects the integrity and personhood of the 'opponent'" and also produces "a better chance of democratic results, . . . because the process itself is more democratic."[13] By treating the other side humanely, nonviolence preserves the hope of a cooperative outcome that sustains rather than strains community.

Most critiques of nonviolence focus on this refusal to use violence, suggesting that it is impractical to take options off the table in an uncertain world. The most common criticism comes from examples that make violence seem like the only option: How could Hitler have been stopped without violence? What would the principle of nonviolence have me do if I see a crazed murderer about to attack an innocent child? There are thoughtful responses to such critiques, but they are beyond my argument.[14] I am not trying to prove here that nonviolence is always the best way to resist violence, only that it is sometimes a viable form of resistance and has something important to teach privileged people who seek climate justice.

As a group, privileged people have not been tempted to use violence to stop climate change, but we have been tempted to ignore the violence caused by our lifestyles and the institutions that support them. Thus, nonviolence can most particularly teach concerned people about the importance of resisting violence and the possibility of doing so without contributing to further violence.

## Pragmatism and Principle

Most accounts of nonviolence distinguish between two different motivations for it: those who are nonviolent for purely pragmatic reasons and those who are nonviolent out of deep moral principle. Gene Sharp has focused his attention on the former category, insisting that nonviolence is a strategic choice, defined "by what people do, not by what they believe." He thus insists that most nonviolent activists choose that path because it makes sense, not out of obedience to some moral rule. He founded the Albert Einstein Institution to train leaders in nonviolence and distributes a handbook for nonviolent struggle, *From Dictatorship to Democracy*, that has been used extensively by protest movements in Eastern Europe, the Middle East, and throughout the world. His core principle is that violence is impractical, which he stresses by using italics: "*By placing confidence in violent means, one has chosen the very type of struggle with which the oppressors nearly always have superiority.*"[15] Nonviolence is a strategic option, to be chosen because it is the most effective way to resist violence.[16]

Other approaches more heavily emphasize moral principle, arguing that nonviolence is an end rather than a means and is worth pursuing whether or not it is effective. For example, Amish communities in the United States are committed to "nonresistance," with a moral commitment that forbids them from serving in the military or the police, raising a hand in anger against another person, or even standing up against opponents in a court of law. An Amish man on trial in 1953 for refusing to serve in the military told the court that "Jesus never killed His enemies. He let his enemies kill Him. . . . Therefore, I'm here to *give myself up* to the jury" (emphasis in the original).[17] Such an approach to nonviolence is principled rather than pragmatic. This Amish man was not motivated by what would happen to him, or even about what would happen to the nation if no one fought for it. He sought to follow the rule laid down in his faith, whatever the consequences.[18]

These examples suggest a choice. Sharp seeks a universal strategy, and so he pays little attention to principles that differ across cultures, philosophies, and religions. By contrast, the Amish separate themselves from mainstream culture with a firm allegiance to their particular beliefs and principles, so they pay little attention to strategic goals that might lead to moral compromise.

One can also find a middle ground that is both strategic and principled. The definition of nonviolence used in this book—a commitment to resist violence while refusing to use violence—attempts to hold pragmatism and principle together, to lift up nonviolence as a response to climate change that does pragmatic good while also exemplifying key moral principles. Such a middle ground, I argue here, is well demonstrated in the tradition of Christian nonviolence.

## CHRISTIAN NONVIOLENCE

Many accounts of Christian nonviolence focus on principles, citing biblical texts that command Christians to avoid violence and emphasizing that faith-based nonviolence tends to be based on moral commitment.[19] This is reasonable because the Christian Bible includes profoundly principled language that seems to command pacifism. In the Gospel of Matthew's Sermon on the Mount, Jesus instructs his audience: "Do not resist an evildoer. But if anyone strikes you on the right cheek, turn the other also" and commands them to "love your enemies and pray for those who persecute you."[20] These biblical verses can easily be interpreted to justify nonviolence; Jesus commands his followers to love their enemies and turn the other cheek, so Christians can say that they are commanded to be nonviolent. However, these instructions are not merely principled; they also lead to very pragmatic movements offering real and effective resistance to violence.

### Loving Enemies

The Sermon on the Mount includes the instruction to "love your enemies and pray for those who persecute you, so that you may be children of your Father in heaven; for he makes his sun rise on the evil and on the good, and sends rain on the righteous and on the unrighteous."[21] For the last two thousand years, Christians have been trying to figure out what this means.

The activist and New Testament scholar Walter Wink insists that the best way to live out Jesus's command is nonviolent resistance. One can only love one's enemies if one refuses to kill or violently harm them, but truly loving them nevertheless means resisting their violent and evil intentions.

Praying for one's enemies means praying and working for their "transformation" and thus requires an approach that can "liberate the oppressed from evil even as it frees the oppressor from sin." In other words, loving one's enemies does not mean simply wishing them well, but also trying to help them to do better, to live less violently. Christians are called to love their enemies "not blindly, but critically, calling them back time and again to their own highest self-professed ideals and identities."[22]

Wink learned about Christian nonviolent resistance in part by spending time in South Africa, where he studied nonviolent protests that came decades after Gandhi and helped to defeat the apartheid system. Under apartheid, black South Africans had separate and inferior educational and health care systems, were forced into segregated neighborhoods, were denied political representation, and were under constant threat of police violence. The victims of this oppression had as much reason as anyone to hate their enemies, and many called for a violent revolution by the majority black population against the minority whites in power. However, some leaders of the resistance against apartheid instead advocated nonviolence and love for their enemies.

After having been jailed for twenty-seven years because of his attempts to overthrow the oppressive government, Nelson Mandela left prison committed to nonviolence, having learned to love his enemies. He writes:

> It was during those long and lonely years that my hunger for the freedom of my own people became a hunger for the freedom of all people, white and black. I knew as well as I knew anything that the oppressor must be liberated just as surely as the oppressed. A man who takes away another man's freedom is a prisoner of hatred; he is locked behind the bars of prejudice and narrow-mindedness. I am not truly free if I am taking away someone else's freedom, just as surely as I am not free when my freedom is taken from me. The oppressed and the oppressor alike are robbed of their humanity.[23]

After apartheid ended, Mandela was elected the first black president of South Africa. In this capacity, he helped to create the Truth and Reconciliation Commission, which sought to help both victims and perpetrators of apartheid violence to process what had occurred and move on to a healed South Africa.

Assisting Mandela in these efforts was the Anglican archbishop of Cape Town, Desmond Tutu, who was also a black South African who had suffered under the injustices of apartheid. Like Mandela, Tutu sought to move away from violence by loving his enemies. "God has no enemies, only family," he insisted, and so "only together, hand in hand, as God's family and not as one another's enemy, can we ever hope to end the vicious cycle of revenge and retribution."[24]

Tutu and Mandela both insisted that if God loves all people, Christians must work to do the same. Their work in South Africa is one example of how a commitment to love one's enemies leads to nonviolent resistance that attempts to heal rather than abolish the enemy, to transform rather than defeat others. This is Christian nonviolence, which seeks not only to follow Jesus's commandments but also to make a positive change by resisting violence.

## Turning the Other Cheek

Of course, as the Amish example above demonstrates, it is possible to understand Jesus's commands as instructions to love one's enemy *without* resistance, even to passively accept violence. This is one interpretation of Jesus's instruction to "not resist an evildoer. But if anyone strikes you on the right cheek, turn the other also."[25] If Christians should "not resist an evildoer," then could active resistance ever be justified? Should Christians passively turn their cheeks rather than take positive action?

Ronald Sider offers a helpful perspective. He notes that many places in Jesus's story show him resisting evil—arguing with mistaken Pharisees, driving moneychangers out of the temple, and standing up to the Roman Empire with his very life. This record of resistance helps Sider notice that the command is not "do not resist evil"; it is "do not resist an evildoer." So, Sider argues, Jesus's words do not forbid his followers from resisting evil, but rather urge them to resist evil while still wishing good for the person doing it. For Sider, this interpretation clarifies that "Jesus's kind of resistance to evil will be of the sort that refuses to exact equal damages for injury suffered."[26] Christians are called to resist evil but to do so without hating their enemies or seeking revenge. The violence of the world must be opposed without using violence.

The notion of "turning the other cheek" might remain puzzling, how-ever, because it is usually interpreted as a passive accepting of violence or even as asking for more abuse. Once again, Walter Wink's analysis is useful. He points out that when it is read in historical context, this instruc-tion is a concrete and clever strategy for resisting violence without using violence. In Jesus's time, to be struck on the right cheek was to be given a backhanded slap, dismissed as an inferior. This was not about causing physical harm but about asserting power. By contrast, to be struck on the left cheek was to be punched, challenged as an equal. Thus, the instruction to turn "the other" cheek was about refusing to be dismissed, refusing to be treated as an inferior. To turn the other cheek was to communicate to one's oppressor, in Wink's words, "I deny you the power to humiliate me. I am a human being just like you. Your status does not alter that fact. You cannot demean me."[27] Thus, in reality, turning the other cheek is not about accept-ing physical violence; it is about resisting the violence of dehumanization.

A twentieth-century example of Christians turning the other cheek comes from the small French village of Le Chambon sur Lignon. During World War II, the town's two pastors were committed to Christian non-violence. So, when Germany invaded France, they instructed their con-gregants not to fight back, not to "resist the evildoer." However, when the Nazis later came into town seeking to register and then to deport the Jew-ish population, the people of Le Chambon refused to cooperate, refused to participate in the dehumanization of their neighbors. Instead, they hid Jewish refugees in schools, orphanages, and the nearby countryside. They helped counterfeiters produce false papers so that these refugees could escape occupied territory. And they filed a petition with the occupied French government opposing the deportation of Jews. These were cre-ative versions of turning the other cheek, of nonviolently insisting on the humanity of all and standing up against an oppressor's violence without using violence. Estimates suggest that between 800 and 5,000 Jews sur-vived because of this single town's refusal to participate in the violence of the Holocaust.[28]

Writing about Jesus's command to turn the other cheek, the lead pastor of Le Chambon, André Trocmé, insists that the key lesson of this verse is about action: "Nonviolence engages evil, it does not withdraw from it," and so Christians are called to act to make God's peace in the world. "Non-violence can only overcome evil if it is the act of God's power on earth, working through human beings."[29] This is a pragmatic, active expression of Christian love that is grounded in deeply held principles.

## FIVE LESSONS OF CHRISTIAN NONVIOLENT RESISTANCE

A few more dimensions of Christian nonviolence can be explored by reintroducing the five witnesses who will guide the rest of this book. Each one demonstrates the richness and complexity of what it has meant for Christians to love their enemies and turn the other cheek. These witnesses show us that nonviolence is creative, structural, courageous, communal, and inclusive.

### Lesson One: The Creativity of Nonviolence

John Woolman lived in the eighteenth century, when most white people in North America simply assumed that the institution of slavery was a part of how the world worked. Slavery was not worth questioning, much less overthrowing. So Woolman had to be creative in his resistance against this violence. When he discovered that clothing dyes were produced by slave labor, he began to wear only undyed clothes, and the sight of his stark white attire called attention to his moral principles. When he dined at others' homes, he refused to eat with silver utensils or plates, because slaves mined silver. Such stark moral commitments were partly about purifying Woolman's own life, but they also served as creative ways to raise awareness about the evils of slavery to everyone he encountered.

Woolman's creativity was also devoted to an empathetic understanding of others. Though he was a white man who had been born free, he learned to imagine the hardships and oppression of slavery, and he encouraged others to do the same. For example, in an abolitionist essay, he asked his white readers to compare their own pains with those of families wrenched apart by the slave trade:

> Our children breaking a bone, getting so bruised that a leg or an arm must be taken off, lost for a few hours, so that we despair of

their being found again, a friend hurt so that he dieth in a day or two—these move us with grief. And did we attend to these scenes in Africa in like manner as if they were transacted in our presence, and sympathize with the Negroes in all their afflictions and miseries as we with our children or friends, we should be more careful to do nothing in any degree helping forward a trade productive of so many and so great calamities.[30]

In other words, if people could imagine themselves suffering the injustices perpetrated by the slave trade, they would never support it. This creative exercise of empathy was an act of nonviolence. Woolman resisted slavery by asking white people to think more broadly, more creatively. At a time when people like him barely thought about the humanity of African slaves, his work to understand their suffering was a form of resistance against dehumanizing violence.[31]

Nonviolence must be creative because it is countercultural—it always works against conventional wisdom and societal norms. The ethicist Ellen Ott Marshall notes that nonviolent activists "envision alternatives" to a culture that seeks to teach them that their enemies are always evil and they are always good, that the only options in the face of violence are fight or flight. The imagination of a nonviolent activist like Woolman helped him to "envision a third way between the two givens of violence and acquiescence" and thus also to "perceive the connection between the two sides of 'us' and 'them.'"[32]

## Lesson Two: The Structures of Nonviolence

In 1889 Jane Addams moved to Chicago's Nineteenth Ward, a poor neighborhood with a population of mostly immigrants, and she remained in residence there across the next four decades. She hoped to empower her neighbors by nurturing their minds and their sense of justice, and she formed deep relationships with these neighbors. She was an educator who cared deeply about personal relationships, but her work was primarily about creating and changing institutions, building systems that resisted violence. Her deepest nonviolence was structural rather than personal.

Addams resisted poverty and racism by creating Hull House, an institution devoted to education and inclusion that continued helping thousands of people long after she died. This same impulse led her to engage other institutions—working for the City of Chicago, lobbying the president of the United States for new laws, and founding an international organization for peace.

Addams never used the phrase "structural violence," but she was profoundly insightful about the power of structures and institutions to shape people's lives and societal assumptions. For example, she criticized union organizers who used what she called "war methods"—siding with one racial group against others or helping union members at the expense of nonunion members. Such methods, she taught, created systemic resentment and hatred, which would be expressed in other parts of the union members' lives. She wrote: "It is fair to hold every institution responsible for the type of man whom it tends to bring to the front, and the type of organization which clings to war methods must, of course, consider it nobler to yield to force than to justice."[33] She built Hull House and many other institutions to demonstrate that "peace methods" could work and to make people better by teaching them how to be less violent.

It is always tempting to personify violence and evil. The stories that are told about the Third Reich in Germany too often focus on Adolf Hitler to the exclusion of the political movement of Nazism and the social forces that gave it power. The stories that are told about the September 11, 2001, terrorist attacks against the United States too often focus on Osama bin Laden rather than the broad movement of Al Qaeda and its political and economic foundations. Such personification is dangerous because it makes evil seem to exist in certain people and nowhere else. But violence is more pernicious, more structural than this. In her struggles against corrupt Chicago politicians, Addams spent little time vilifying their greed or arrogance but instead tried to figure out how the system was structured to reward greed and arrogance and how this system could be changed.[34] She sought to defeat not evil people but evil structures. Her conception of nonviolence was systemic.

As the activist and scholar Walter Wink puts it, "Evil is not just personal but structural and spiritual. It is not simply the result of human actions, but the consequence of huge systems over which no individual has full control."[35] Institutions shape human beings, so making human beings less violent requires structural changes to these institutions.

## Lesson Three: The Courage of Nonviolence

Dorothy Day, who devoted her life to serving alongside the poor in New York City and witnessing for peace, took a resolutely nonviolent stance against US involvement in World War II. She believed that Christians are called to love their enemies and that such love could not be compatible with war. This was a profoundly unpopular position, and therefore, as the war began, many who had joined and supported her work on behalf of the poor turned away because they did not want to be associated with her stance.

Day remained resolute in her commitment. However, it bothered her when critics called pacifists "cowardly," implying that those who resisted war were motivated by a fear of violence. She defended herself and others in her movement by noting that they had made the brave decision to live among the poor and that by living with the poor they saw the depths of violence and oppression far more deeply than most: "Let those who talk of softness, of sentimentality, come to live with us in the cold, unheated houses of the slums. Let them come to live with the criminal, the unbalanced, the drunken, the degraded, the perverted. . . . Let them live with rats, with vermin, bedbugs, roaches, lice."[36] She insisted that no one who had chosen to live in the poorest part of New York City could possibly be dismissed as a coward.

It is certainly possible to argue that Day was wrong to oppose military action in World War II, but it is unfair to argue that she was a coward who feared conflict. Throughout her life, she subjected herself to dangerous conditions in order to be with the poor. In World War II and at many other times, she took an unpopular pacifist stance and lived with the consequences. Her commitment to nonviolence was brave precisely because it was countercultural.

To be consistent and effective, nonviolence must be courageous—it must show a willingness to endure hardship and criticism, over months, years, or a lifetime. Ronald Sider notes that Christian nonviolence is based upon the model of Jesus, who was crucified for his resistance against hatred, oppression, and the Roman Empire. For those who believe in it,

Sider writes, "the cross is not some abstract symbol of nonviolence. The cross is the jagged slab of wood to which Roman soldiers spiked Jesus of Nazareth whom we follow and worship."[37] Christian nonviolence requires courage because it leads to real pain and marginalization, as it did for Jesus. Dorothy Day took courage from the story of Jesus, a story that helped her to live with rats and bedbugs and the harsh judgment of many who dismissed her.

## Lesson Four: The Community of Nonviolence

Christian nonviolence can also lead to death, as it did for Martin Luther King Jr. During his very first organized protest in Montgomery, Alabama, racists set his house on fire while his wife and daughter were inside. They survived, but King knew from this moment that his resistance against the violence of racism, poverty, and militarism could be costly. In his final sermon, preached in Memphis the night before he was shot, he spoke about this cost. He told the story of another assassination attempt, when he was stabbed while signing books. He noted that every plane he boarded had to be searched an extra time because of the likelihood that someone had placed a bomb on it. He knew his death was possible. Facing this fact, he told his audience,

> I don't know what will happen now. We've got some difficult days ahead. But it doesn't matter with me now. Because I've been to the mountaintop. And I don't mind. Like anybody, I would like to live a long life. Longevity has its place. But I'm not concerned about that now. I just want to do God's will. And He's allowed me to go up to the mountain. And I've looked over. And I've seen the promised land. I may not get there with you. But I want you to know tonight, that we, as a people will get to the promised land. And I'm happy, tonight. I'm not worried about anything. I'm not fearing any man.[38]

These words testify to King's deep courage and his profound faith, but they also show that he understood the movement for nonviolent justice as something far bigger than himself. He was not concerned about what would happen to him but rather what would happen to the community.

Nonviolence can only work if it is undertaken alongside other people who can prop up one's courage, help to tackle violent structures, and inspire creativity. Five years before his death, writing about the movement against segregation in Birmingham, King noted that change was accomplished not by his speeches but by the community of "sit-inners and demonstrators," "young high school and college students," and "old, oppressed, battered Negro women," such as the one who encouraged everyone to sustain an exhausting protest with the words "my feets is tired, but my soul is at rest."[39]

Martin Luther King Jr. was an amazing human being, as were the other four witnesses discussed in this book. But none of them worked alone; all were part of movements. All participated in, helped to nurture, and left behind communities of nonviolent resistance.

## Lesson Five: The Inclusivity of Nonviolence

In 1962 Cesar Chavez, Dolores Huerta, and a community of farmworkers formed a union to demand better working conditions, better pay, and basic human rights. As the leader of this union for the next thirty years, Chavez learned that he could best serve farmworkers if he made connections to other communities beyond them. To fund national strikes, he had to build partnerships with other unions, churches, and donors who could help support his workers and their families. Before the grape growers would take his demands seriously, he had to mobilize a national boycott of their products. As he studied the health effects of pesticides on his union's members, he also learned about the effects on their children, on the people who bought and ate the food, and on the natural world. His nonviolent movement was inclusive because he believed he could only resist violence against farmworkers by working to protect all people and all creatures.

In contrast to the other four witnesses, it is appropriate to call Chavez an environmentalist because he wrote and thought a great deal about how his resistance against violence should advocate justice not only for the entire human race but also for other creatures. This influenced his personal discipline, as well, and he was a dedicated vegan and animal rights activist. When presented with an award from In Defense of Animals, he told them: "The basis for peace is respecting all creatures. . . .

We cannot hope to have peace until we respect everyone, respect ourselves, and respect animals and all living things."[40] This broadly inclusive approach to nonviolence is vital for climate justice, which requires that people resist the global violence of climate change against every person and every living thing.

The same idea has been expressed more recently by the Catholic peace activist Brayton Shanley, who insists that "nonviolent peace activism stands in opposition to our society's plague of violence and cannot simply remain in the historically limited framework of anthropocentrism—that is, human beings mediating their conflicts on earth with regard only to other human beings."[41] If nonviolence is to truly challenge the structures of violence, it must challenge the structures that separate humanity from the rest of the world. If nonviolence is to creatively build communities, it must extend these communities beyond any single species. Nonviolence is radically inclusive.

## RESISTING THE VIOLENCE OF CLIMATE CHANGE

The violence of climate change will not be solved, and there is no single response that will be sufficient or satisfying by itself. However, this does not mean that the problem is hopeless. Much to the contrary, it means that many responses can and should be tried. These responses should be inclusive, because climate change threatens every creature on the globe. Responses to climate change should be communal, because no one can understand or meaningfully resist this violence alone. Responses to climate change should be courageous, because it is tempting to be overwhelmed and paralyzed by fear in the face of a problem so large and so complicated. Responses to climate change should be structural, because destructive institutions and systems have developed and solidified over time. Finally, responses to climate change should be creative, because the adaptive complexity of this violence exceeds conventional wisdom.

Nonviolence offers a countercultural tradition of creativity, structural thinking, courageous action, communal partnerships, and inclusive attention. This tradition will not solve climate change, but it will help people to resist its violence while learning what it would take to stop contributing to further violence. We can learn more about such resistance from the five witnesses to whom we now turn.

## NOTES

1. James, "Moral Equivalent of War," 353, 352, 356.
2. Ibid., 359.
3. See especially Gandhi, *Essential Gandhi*; Gandhi, *Autobiography*.
4. The journalist Mark Kurlansky notes that there is no positive word for "nonvio-lence." That it can be discussed only as a negation of something else signals how revolutionary an idea it is—"an idea that seeks to change the nature of society, a threat to the established order." Kurlansky, *Nonviolence*, 5.
5. For multifaith considerations of nonviolence, see especially Smith-Christopher, *Subverting Hatred*. For a powerful justification of nonviolence that does not depend upon any religious tradition, see especially Deming, *Part of One Another*.
6. The social ethicist Reinhold Neibuhr, who argued emphatically against pacifism for most of his public life, made a point of noting that he still believed he had much to learn from pacifists. In an essay advocating military engagement before the United States had joined the allies in World War II, for example, he wrote: "If there are men who declare that, no matter what the consequences, they cannot bring themselves to participate in this slaughter, the Church ought to be able to say to the general community: We quite understand the scruple and we respect it. It proceeds from the conviction that the true end of man is brotherhood, and that love is the law of life. We who allow ourselves to become engaged in war need this testimony of the absolutist against us, lest we accept the warfare of the world as normative, lest we become callous to the horror of war, and lest we forget the ambiguity of our own actions and motives and the risk we run of achieving no permanent good from this momentary anarchy in which we are involved." Niebuhr, "Why the Church Is Not Pacifist," 146.
7. For histories of nonviolence as an idea and movement, see Juhnke and Hunter, *Missing Peace*; and Chernus, *American Nonviolence*. For case studies on nonvi-olent struggles, see Ackerman and Kruegler, *Strategic Nonviolent Conflict*; and Ackerman and DuVall, *Force More Powerful*.
8. Sharp, *Politics of Nonviolent Action, Part One*, 64.
9. Sharp, *Politics of Nonviolent Action, Part Two*.
10. Gandhi, *All Men Are Brothers*, 98.
11. Sharp and Paulson, *Waging Nonviolent Struggle*, 27.
12. David Hoekema makes this point well, writing that "realism about human nature . . . undermines the assumption that weapons of destruction and violence

intended to restrain evil will be used only for that purpose. The reality of human sinfulness means that the instruments we intend to use for good are certain to be turned to evil purposes as well. There is therefore a strong presumption for using those means of justice that are least likely to be abused and least likely to cause irrevocable harm when they are abused. An army trained and equipped for national defense can quickly become an army of conquest or a tool of repression in the hands of an unprincipled leader." Hoekema, "Practical Christian Pacifism," 918–19.

13. Sider, *Nonviolent Action*, xv, 159–60.
14. See especially Wink, *Engaging the Powers*, chap. 12; and Brimlow, *What About Hitler?*
15. Gene Sharp, "From Dictatorship to Democracy," 4.
16. Other explorations of nonviolence that focus on a pragmatic approach include those by Boserup and Mack, *War Without Weapons*; Ackerman and DuVall, *Force More Powerful*; and Schock, *Unarmed Insurrections*.
17. Kraybill, Nolt, and Weaver-Zercher, *Amish Way*, 5–6. For a moving account of the kind of forgiveness that such nonresistance makes possible, see the same authors' *Amish Grace*.
18. Other explorations of nonviolence that focus on moral principle include those by Yoder, *Politics of Jesus*; Kurlansky, *Nonviolence*; and Cortright, *Gandhi and Beyond*.
19. For collections of Christian writings about nonviolence that offer far more comprehensive and detailed accounts of the tradition than I can here, see especially O'Gorman, *Universe Bends toward Justice*; and Long, *Christian Peace and Nonviolence*. For discussions of nonviolence amid other Christian responses to violence, see especially Allen, *War*; Cahill, *Love Your Enemies*; and Allman, *Who Would Jesus Kill?*
20. Matthew 5:9, 39, 44.
21. Matthew 5:44–45.
22. Wink, *Powers That Be*, 110–11; also see 34.
23. Mandela, *Long Walk to Freedom*, 544.
24. Tutu, *God Has a Dream*, 47, 58. Tutu has more recently called for a divestment campaign from fossil fuels modeled on the divestment campaign that helped to weaken the apartheid South Africa government: "Just as we argued in the 1980s that those who conducted business with apartheid South Africa were aiding and abetting an immoral system, we can say that nobody should profit from the rising temperatures, seas and human suffering caused by the burning of fossil fuels." Desmond Tutu, "We Fought Apartheid, Now Climate Change Is Our Global Enemy," *Guardian*, September 20, 2014, www.theguardian.com/commentisfree/2014/sep/21/desmond-tutu-climate-change-is-the-global-enemy.

25. Matthew 5:39.

26. Sider, *Christ and Violence*, 47–48.

27. Wink, *Engaging the Powers*, 176.

28. See especially Hallie, *Lest Innocent Blood Be Shed*; and Sauvage, *Weapons of the Spirit*.

29. Trocmé, *Jesus and Nonviolent Revolution*, 153.

30. Woolman, *Journal and Major Essays*, 233.

31. Laura Hartman argues that such creative imagination is a crucial gift that Woolman gives to morally serious people in our time. See Hartman, *Christian Consumer*, 178.

32. Marshall, "Practicing Imagination," 66.

33. Addams, *Newer Ideals*, 80.

34. See especially Addams, "Why the Ward Boss Rules," in *Jane Addams Reader*, 118–24.

35. Wink, *Powers That Be*, 31.

36. Day, *By Little and By Little*, 263.

37. Sider, *Christ and Violence*, 16.

38. King, *Testament of Hope*, 263.

39. King, *Why We Can't Wait*, 94.

40. Quoted by Elliot M. Katz, "Cesar Chavez: A True Guardian," March 27, 2013, www.idausa.org/remembering-cesar-chavez/.

41. Shanley, *Many Sides of Peace*, 12–13.

# PART II

Five Witnesses of
Nonviolent Resistance

# 3

## John Woolman's Moral Purity and Its Limits

The language of Christ is pure, and to the pure
in heart this pure language is intelligible; but in
the love of money the mind being intent on gain
is too full of human contrivance to attend to it.
—John Woolman, *Journal and Major Essays*

Because John Woolman was a literate merchant in the eighteenth century, when many could not read or write, one of his duties in his early twenties was to prepare legal documents. But when his employer asked him to write a bill of sale for a slave, Woolman felt troubled by the idea of "writing an instrument of slavery for one of my fellow creatures." Bound by duty, he wrote the bill after telling his employer and the customer of his reservations, but he remained uncomfortable. The next time he was asked to write such a document, he listened to his conscience rather than his contract and refused. He was surprised to find that, like him, the man buying the slave "was not altogether agreeable in his mind" about slavery. In this instance, and in many others, Woolman's commitment to live out his principles helped others to recognize and hold themselves accountable to their own.[1]

Woolman offered a strident moral witness in the face of structural violence, standing up to poverty, to injustice toward Native Americans, and, most of all, to the horrors of slavery. He firmly insisted that good people are called not only to oppose violence in all its forms but also to remove themselves from the systems that make them complicit in violence.

Woolman's witness was motivated by his membership in the Society of Friends, commonly known as the Quakers, and most of his writings are directed to fellow Quakers. In the quotation that begins this chapter, he suggests to this audience that the only justifications white people had for slavery came from self-serving greed.[2] In his time, this would have been a controversial and shocking statement, as many white audiences simply assumed slavery as a norm. But Woolman insisted that if his fellow Quakers purified their hearts by removing greed, they would see the obvious truth that no human being should be treated like the property of another. He believed that anyone who tried could purify their heart, that all who worked to hear it would understand Jesus Christ's "pure language." Woolman tirelessly worked to live a life that would demonstrate this, and his witness of nonviolence and moral commitment has now endured two and half centuries.

It is important to distinguish between the structural violence of slavery and the structural violence of climate change, between the challenges of the eighteenth century and the twenty-first. But Woolman is nevertheless a fitting witness with whom to continue this exploration of how contemporary concerned people should respond to the violence of climate change. His own activism was so personal—he simply refused to participate in systemic violence—that he has much to teach those who seek to resist the violence of burning fossil fuels, deforestation, and industrial food production. With Woolman in mind, we can begin to think about what it could mean to resist violence in our own lives. In dialogue with Woolman, we can begin to reflect on the power and limits of such a personal witness.

## SEEKING A SINCERE HEART

John Woolman's legacy is most fully preserved in the journal he wrote about his life's moral struggles. One of the first stories he recounts therein concerns an act of violence. Passing the time as a young boy in rural New Jersey when it was still a British colony, he encountered a mother robin nervously flying around her nest to protect her chicks. Seeing a challenge, he threw a stone and hit the bird squarely, striking her dead. He writes: "At first I was pleased with the exploit, but after a few minutes was seized with horror, as having in a sportive way killed an innocent creature while she was careful for her young." He thought about the young birds he had just orphaned, and realized that they would slowly die of starvation. Then, "after some painful consideration on the subject, I climbed up the tree, took all the birds and killed them, supposing that better than to leave them to pine away and die miserably."[3]

Writing years later, Woolman concludes this story in two ways. First, he reports that it disturbed him deeply, that for some time after he "could think of little else but the cruelties I had committed, and was much troubled." But he also draws a larger conclusion from the experience, pondering what it was inside him that drove him to question his own behavior and lament its consequences. He concludes that every human being has an internal tendency for "goodness toward every living creature," which teaches us to be "tender-hearted and sympathizing."[4] He had ignored this internal tendency when he killed the mother bird, just as he later ignored the goodness within himself when he wrote a bill of sale for a slave.

Woolman is hardly the only child to commit a careless act of cruelty, but he is admirable for his willingness to tell the story and for the ways he continued to reflect upon how to resist the violence of which he knew he was capable. Indeed, in many ways his public life can be understood as an attempt to follow the best impulses within himself, seeking to avoid feeling "much troubled" by his conscience.

He believed that human beings are by nature kind, peaceful, and caring, and he sought to remove and resist obstacles that prevented people from acting on these impulses. This meant avoiding and warning others about the distractions of wealth and worldly success, insisting that the native peoples of North America had rights and should be treated fairly, expressing concern for the well-being of whales and horses, and making passionate arguments against the institution of slavery, which he knew was stifling the lives of enslaved persons and warping the souls of slave owners.

Woolman was born in 1720 as the eldest of fourteen children, and custom suggested that he would take over his father's large farming estate. However, at the age of twenty-one years, he felt called to a simpler life and asked his father's permission to instead work as a merchant. He found success in that work but eventually began to look for an even simpler job that would allow him more mobility and flexibility. This led him to tailoring, which supported him, his wife, and their daughter for the rest of his life.

Woolman's choice of a simple profession demonstrates his belief that "a humble man with the blessing of the Lord might live on a little," and that further material success would distract him from the work of his faith.[5] The historian Michael Birkel notes that Woolman exemplifies traditional Quaker testimonies, which call believers to live simply, equitably, and peacefully. Woolman saw that "failure to lead a simple life ultimately leads to oppression of others, and this in turn can lead to war." This is not merely a social analysis but also a spiritual one; Woolman believed that satisfaction in life comes not from wealth or reputation but from opening "ourselves fully to the love of God."[6]

Above all else, Woolman was unflaggingly principled. As discussed above, he refused to write a bill of sale for another human being while he was a merchant. Later in his life, when he learned that clothing dyes were produced in part through the labor of slaves, he decided to wear only white, undyed clothes, sending a clear signal of resistance to everyone he met.

Woolman's open discussion of his own moral development and challenges was part of his abolitionist strategy. He shared his own uneasiness with the institution of slavery in order to allow others to express theirs. Because he believed all people have goodness within them, he offered opportunities for his neighbors to understand and articulate the violence of slavery. Thus, his first antislavery essay begins by observing his own "real sadness" about the violence of the institution and its fundamental incompatibility with "an enlightened Christian country." This issue had "been like undigested matter in my mind," and so Woolman felt called to share his struggle.[7]

For thirty years, Woolman not only wrote in opposition to slavery but also traveled to Quaker communities across the North American colonies and in England to share his ideas in person. Because the Society of Friends was so deeply committed to peace, he frequently pointed out that human beings had been made into slaves by the violence of war, kidnapping, and imprisonment.[8] To those who suggested that slavery was a necessary part of economic structures, he insisted that everything on Earth ultimately belongs to God, and so economic structures should serve moral principles rather than the other way around.[9] To those who suggested that people of African descent were inherently inferior to Europeans, he insisted that God had created everyone, and so all people are imbued with the same internal goodness.[10]

As demonstrated by the story at the beginning of this chapter, Woolman was committed to acting on his beliefs in addition to talking about them. And other Quakers learned from this commitment.[11] One story—not told in his own journals—is that he once came to dinner at the house of a wealthy Quaker but, as soon as he learned that this man owned slaves, quietly left. This was not simply a retreat but also the most effective way to communicate his moral protest, and the theologian Elton Trueblood reports that the slave owner felt the power of Woolman's moral witness and decided to liberate his slaves the next day.[12]

Woolman did not frequently sneak out of others' homes, but when he ate or stayed with slave owners, he insisted on paying any person who had helped to care for him. His commitment to avoid dyed clothes put him soundly against the fashions of his time, and he stood out as, in his own word, "singular." His appearance invited conversation about his moral principles.[13]

Such singular behavior was not restricted to Woolman's abolitionism. At a time when most of his compatriots dismissed or oppressed Native Americans, he made a point of advocating for their fair treatment, and he undertook missionary work among natives with a commitment to recognizing the goodness inherent in them rather than converting them away from their traditions. When he booked passage to England, he took time to berate the ship's Quaker captain for all the ornate decorations of the boat, the cost of which could have gone to feed the hungry. He then insisted on staying in the least comfortable lodgings on the boat in order to avoid luxury and wasteful expense and to better understand the plight of the poor. Once he arrived in England, he was appalled at the way horses were treated, reporting that they were forced to run so far so quickly that some of them died or went blind in the course of business. So he insisted on walking everywhere, and he asked his family not to send him letters by post so that no horses would be punished for his sake.

Woolman died of smallpox in England in 1772, soon after having convinced a Friends Meeting in London to publicly renounce slavery. His *Journal* was published two years after his death, and it has remained in print ever since as the testament of an early and influential abolitionist, a deeply spiritual quest, and an individual seeking to live rightly in a world of profound structural violence.

## "THAT PURITY WHICH
## IS WITHOUT BEGINNING"

John Woolman's life was spent avoiding the mistakes he saw in his culture, seeking the pure voice of goodness inside himself by separating it from the competing voices of selfishness, acquisitiveness, and slavery.[14] At the core of his resistance was a faith in human goodness. He could move away from the assumptions and mistakes of his society because he believed that there was a presence and light within him—and within all people—offering access to deeper truths. As he put it: "There is a principle which is pure, placed in the human mind, which in different places and ages hath had different names. It is, however, pure and proceeds from God. It is deep and inward, confined to no forms of religion nor excluded from any, where the heart stands in perfect sincerity."[15] This pure principle is deeper than the customs of human cultures, bigger than any religious institution, truer than anything made by human beings.

Woolman was devoutly Christian—fundamentally shaped by his reading of the Bible, by the traditions of the Society of Friends, and by a belief in the God represented by the teachings of Jesus Christ. However, he was also critical of anyone who professed belief "in one Almighty Creator and in his son Jesus Christ" but did not demonstrate this belief by treating other people well. Indeed, he was confident that anyone "established in the true principle of virtue" could be counted among those "who fear God and work righteousness," whether or not they were Christian. His concern was not for doctrinal purity but for moral purification; his strongest desire was not to convert people to Quakerism but rather to convert them to a life lived according to the "perfect sincerity" to be found in every human heart.

Woolman prayed for faithfulness to this sincerity before talking with slave owners, asking God "that I might attend with singleness of heart to the voice of the True Shepherd and be so supported as to remain unmoved at the faces of men."[16] He knew that he had a selfish desire to tell people what they wanted to hear, and so he prayed for the courage to speak the truth that slave owners needed to hear rather than the lies they wanted to believe. His responsibility to other people was not to indulge them but to show them what a truly moral life looked like. He sought purification not only for his own sake but also for his neighbors', and he called all people of goodwill to do the same, to become "a light by which others may be instrumentally helped on their way, in the true harmonious walking."[17]

The goal of purification implies that there is an impurity, some confusion or contamination that needs to be removed from the soul. For Woolman, this impurity is best understood as selfishness. The desire to build one's self up through wealth or popularity leads people to compromise their morals. Such selfishness was, for Woolman, the root of slavery. This incredibly violent system gradually made its way into European culture, and it began to seem normal because it provided economic benefits. As it became impolite to question the system of slavery, people began to ignore their natural questions and doubts about treating other humans as property. Woolman writes, "When self-love presides in our minds . . . by long custom, the mind becomes reconciled with it and the judgment itself infected."[18] In other words, selfish human beings can easily become so accustomed to violence that they do not talk about or resist it, that they stop even noticing it.

Many slave owners resisted Woolman's arguments by suggesting that they had a responsibility to their children, to pass on their slaves as wealth for the next generation. Woolman also pointed out the profound selfishness of this thinking. He argued that this approach is not only cruel to the slaves who deserve freedom but also, ultimately, to the children of slave owners, who are raised to think they should have power over other human beings and warped away from "that humility and meekness in which alone lasting happiness can be enjoyed."[19] As children of European descent were raised to believe that slavery is acceptable and children of African descent were raised without rights and liberties, Woolman saw "dark matter gathering into clouds over us." He correctly predicted that slavery was embedding violence and oppression deep into the structures of the North American colonies.[20]

For Woolman, the moral response to such violence was clear: resistance. This involved, first, an unwillingness to conform to the mainstream society around him, a commitment "to be a fool as to worldly wisdom."[21] Whereas other parents worked to leave large inheritances, he celebrated the chance to leave behind "little else but wise instructions, a good example, and the knowledge for some honest employment."[22] Although other Christians rationalized slavery using biblical verses and cultural traditions, he insisted that all human beings are equal in the eyes of God and so should have equal liberty. And though other abolitionists were quietly polite around slave-owning peers, Woolman refused to compromise and made his principles known.

Woolman's witness demonstrates the power of testing every aspect of life against the "perfect sincerity" deep within one's own heart, resisting the personal temptation of passing desires or the social temptation to accommodate other people with easy answers. Self-discipline and self-examination give people the freedom to follow conscience rather than custom, to show others the guidance of purity by dismissing the infected judgments of selfishness.

## CLIMATE CHANGE AND MORAL PURIFICATION

A deep challenge of climate change is that all people in the industrialized world are part of the problem. We use fossil fuels, eat industrialized food, and participate in global commerce. As one small example, substantial carbon dioxide ($CO_2$) has been released in the process of getting this book into your hands. $CO_2$ was burned to manufacture and power the computer on which it was written, the machines that bound it between covers or downloaded it onto e-readers, and the machines that got the object you are holding into your hands. Contemporary media are utterly dependent upon fossil fuels and industrial systems, so taking a public stand against the violence of climate change creates moral tension between medium and message.

Those who deemphasize the moral importance of climate change frequently use this moral tension as a basis for criticism. When former vice president Al Gore released his film *An Inconvenient Truth*, many critics decried the fact that he frequently flies, often in private jets, to give speeches about the dangers of $CO_2$ emissions. Jet planes are incredibly damaging to the climate, and a private jet places the responsibility on one or few people rather than spreading it across many passengers. Gore also came under criticism for the luxuriousness of his house. The day after his film won an Academy Award, a conservative think tank said that he deserved "a gold statue for hypocrisy" because, at that time, his residence annually consumed more than twenty times more energy than that of the average US house.[23]

The philosopher Brian Henning offers a corrective nuance to such criticisms, distinguishing between hypocrisy and what he calls "moral finitude." Hypocrisy comes when someone contradicts their own values without regret or any attempt at correction, accepting the gap between moral ideals and moral actions without concern. Finitude, conversely, recognizes that people live in social contexts and structures that make it impossible for most of them to live up to all their highest moral ideals. So Henning suggests that it is not useful to label everyone who falls short of every moral goal a hypocrite: "Rather, those who earnestly pursue but fail to fully achieve their moral ideals are morally finite." Henning recommends that people devote their energy less to pointing out others' imperfections and inconsistencies and more to finding "the courage to declare, scrutinize, and pursue" one's own ideals to the best of one's ability.[24]

Given the reality of moral finitude, the question for those of us who are concerned about climate change is not how to avoid all complicity in the problem. Such purity is impossible. But we should nevertheless work to declare, scrutinize, and pursue a life that will cause less harm to the world, to our human and nonhuman neighbors, and to future generations. The moral challenge is to do less damage to the climate and to continually struggle to change the systems that lead every participant in industrial civilization to cause such damage. The goal is purification, a movement toward purity that recognizes human finitude.

## The Quest for Purification

Some activists have taken extreme measures to purify their lives in light of climate change. The meteorologist Eric Holthaus responded to the release of a 2013 report by the Intergovernmental Panel on Climate Change (IPCC) with a vow to never again fly in a plane, tweeting that his travel "is not worth the climate." The IPCC report called for "substantial and sustained reductions" in $CO_2$ emissions, and Holthaus calculated that he and his wife had caused 33.5 metric tons of the gas to be released by flying in the previous year. He wrote, "I've begun to heed the IPCC's call to action. Individual gestures, repeated by millions of people, could make a huge difference."[25]

Forty years earlier, John Francis made an even more radical break from the fossil fuel economy when he stopped using any motorized transportation in response to a 1972 oil spill in San Francisco Bay. He began to walk everywhere he went. This led to "almost constant bickering and arguments with my friends as to the question of whether one person walking could make a difference." To stop these arguments, Francis made another drastic commitment, taking a vow of silence so that he could fully hear others' voices without defending his own position. He spent twenty-two years traveling around without motorized vehicles and seventeen years not speaking. During this time he built a sailboat, walked across the continent, and earned bachelor's, master's, and doctoral degrees. His commitments began as a response to environmental pollution, but he learned a lesson that captures the interconnected nature of structural violence; the "environment" he sought to protect is not only the San Francisco Bay and the atmosphere, it "is also about human and civil rights, economic equity, gender equality, and from the standpoint of a pilgrim on the road, environment is about how we treat each other when we meet each other."[26]

Neither Holthaus nor Francis has made a dramatic impact on the quantity of climate-changing gasses in the atmosphere. Each is only one person, with a finite impact constrained by the systems in which they live. However, both made dramatic changes in their own lives and by doing so influenced the people they met. Along these lines, Francis offers helpful advice to others who seek to change the world: "The only person one has the ethical authority to change is oneself. When we change our self, we indeed change the world. As we continue our journey, we can make a difference in our community and in the world, one step at a time."[27] For seventeen years, every person who tried to speak with Francis learned from his silence. Similarly, every person who asks Holthaus to travel learns about his commitment and, through it, about the seriousness of climate change.

These two activists represent one way to push against moral finitude. They reduce their complicity in the fossil fuel economy and, with their example, hope to inspire others to do the same. This is a response to climate change that begins by seeking first to purify oneself.

## The Limits of Individual Action

The philosopher Baird Callicott points out the limitations of such individual purification. He reports that when he first began to learn about environmental problems, he was "Joe Bioregionalist." He limited driving and travel, grew his own food, and turned his home heating system down—even through cold Wisconsin winters. This behavior, similar to that of John Francis and Eric Holthaus, was Callicott's attempt to "do his bit" and show his neighbors what a sustainable life might look like. However, as he was working to live locally, "the US economy kept growing, and most everyone else went on consuming fossil fuels. . . . Industrial agriculture expanded. More consumer items than ever came on the market." It is not enough for everyone to do their bit, Callicott eventually concluded, because each person's actions are "but an insignificant drop in a very leaky bucket."[28]

For Callicott, environmental degradation cannot be fixed by individual action because it is not an individual problem. It is, instead, a collective problem caused by the society as a whole and rooted in the basic worldview and beliefs that inform it. So he now devotes his energies to education, challenging the dominant worldview and the beliefs it has inspired. He writes to help his readers "come to see nature as a systemic whole and ourselves as thoroughly embedded in it, a part of nature, not set apart from it." He is confident that when people truly believe this, their politics and their individual behavior will naturally change.[29] The vital task is to reach others rather than to purify oneself. If buying processed food and flying in planes gives Callicott more time to write and a broader platform from which to speak about environmental problems like climate change, he is prepared to do so.

Martin Palmer, the head of the international Alliance of Religions and Conservation, offers another critique of environmental self-purification, which he critically labels "Neo-Puritanism." He recalls meeting a scientist so devoted to environmental issues that he and his wife refused to fly in planes, even though this commitment meant they had missed both their sons' weddings and never met their grandchildren. Palmer laments that this man had forgotten "how to party" and suggests that such a puritanical approach to the environment is never going to convert people to the cause. Indeed, he worries that this approach will, instead, discourage others from taking the issue seriously. It will be a challenge to recruit more people to show concern about climate change if such concern seems to require disconnection from loved ones and a refusal to party. Therefore, Palmer argues that the role of religion in a time of environmental crisis should not be to take away pleasures but to remind people how to celebrate one another and the Earth.[30]

### Struggling with Moral Finitude

John Francis's and Eric Holthaus's attempts to separate themselves as much as possible from environmental destruction represent one response to moral finitude. Martin Palmer and Baird Callicott offer a different response, appealing to a broad audience in order to spread the word about the challenge and the importance of developing united, large-scale solutions.

Privileged people called to respond to the violence of climate change must take these options seriously and carefully discern how to balance them—how to weigh the call for individual purification against the need to relate and appeal to the surrounding culture. Climate justice requires courageous people to act, but it cannot be solved by individual actions. The moral challenge of striking this balance calls for careful discernment in conversation with John Woolman's witness.

## PERSONAL PURIFICATION
## AS A POLITICAL ACT

John Woolman believed that individuals can change the world by modeling an alternative to selfishness and violence. The inner voice inside every person seeks to be good, and so the task of an activist is to demonstrate that such goodness is possible.

The philosopher Phillips Moulton explains that Woolman's effectiveness came from his "ethical purity," the integrity that made a profound impact on everyone he met.[31] The theologian Sallie McFague calls Woolman "a walking parable" because he so consistently challenged others to reexamine their own behaviors through his personal commitment.[32] He wore undyed clothes not only to keep himself free from the impurity of structural violence but also to call others' attention to that violence. He insisted on sleeping in the least comfortable quarters on a transatlantic ship not only to cleanse himself of luxurious indulgences but also to learn about the plight of those who had no other option.

Of course, Woolman's integrity and purity did not end slavery, poverty, or oppression. But he did convince some people to free their slaves, to give away their wealth, and to listen more closely to their own best impulses. And he helped to energize an abolitionist movement that continued for almost a hundred years after his death, until his country finally made that vile institution illegal. Concerned people who seek to oppose the structural violence of climate change in the twenty-first century have much to learn from this eccentric eighteenth-century tailor—because the systems of violence are related, Woolman's ability to speak out against slavery was singularly powerful, and we, too, should purify ourselves from the burdens of structural violence.

## Slavery and Climate Change

Slavery is a unique horror in human history. Treating a human being as property is a fundamental betrayal of the most basic commitments in human civilization, and the torture and degradation inflicted upon slaves in the United States was unconscionable. Fortunately, the chattel slavery that John Woolman opposed is today illegal and is widely understood as morally horrific. Sadly, however, slavery continues in the contemporary world. The Quaker activist Kevin Bales estimates in his 2012 book *Disposable People* that more than 27 million human beings are currently enslaved in some way.[33] Woolman's witness is vital for those who struggle against this crime.

Woolman also offers resources to those who would oppose other forms of structural violence, and he repeatedly emphasized that every moral challenge has a common root.[34] He saw selfishness and greed as the source of all social problems; slavery, war, and mistreatment of Native Americans are all expressions of the same violence. All emerge from the desire of some to have more than they need at the expense of others. Indeed, centuries before the contemporary environmental movement, he noted that the same greed that sent soldiers to war and sailors into perilous seas led to excessive logging and whale hunting, perpetrating violence on both the nonhuman environment and on human beings.[35]

Woolman offers no answer to the contemporary problem of climate change; his world and today's are different in vitally important ways. Nevertheless, there are connections, and the violence he opposed is connected to the violence faced by concerned people in the twenty-first century. Along these lines, the climate activist David Orr suggests that future generations will judge the people of the early twenty-first century "much as we now judge the parties in the debate prior to the Civil War" who made lengthy economic arguments about slavery but ignored its moral dimensions. Slavery and climate change both "inflate wealth of some by robbing others," and both "warp and corrupt politics and culture."[36]

Michael Birkel makes the same analogy on a more personal level: "Slaveholders knew that slavery was not really justifiable morally, just as we know today that degrading the environment, for example, is inexcusable. . . . The truth gnaws at our comfort; we are not happy. When we suppress the light of God within us, we need to be redeemed from this situation."[37] Birkel goes on to argue that Woolman provides the resources to work toward such redemption. Two such resources particularly relevant to the struggle against climate change are Woolman's awareness of his own limits and his cultivation of deep empathy for all involved in structural violence.

## Purification Rather Than Purity

John Woolman is a challenging witness from whom to learn because he was so uncompromising. In a letter introducing him to Quakers in England, a colleague noted that Woolman followed "a straighter path than some other good folks are led or do travel in," and advised his hosts that because Woolman would "do nothing knowingly against the Truth," "little will content him."[38] Translated into contemporary terms, Woolman's colleague was warning the British that this man would be a handful; he did not know how to party. He was the kind of person who might ask his family to stop sending him mail because of the ways horses are treated, who was liable to start a moral discussion when you just want him to draw up a contract. He was a strange person by the standards of his time—or any other.[39]

Those who seek to learn from Woolman must be aware that their resistance to the violence of climate change, too, may seem strange. John Francis's vow of silence surely seemed strange. Eric Holthaus's decision never to fly on an airplane certainly seems strange, at least to wealthy and upper-middle-class people in the industrialized world, the small portion of the global population who assume that they should have the freedom to traverse long distances for both business and vacations. These are risky, controversial choices. Indeed, to follow Woolman is to embrace the strangeness of these acts as part of the point. His integrity was powerful because it stood out, it set an example that others could not help but notice. Today's communities will have a harder time ignoring the structural violence inherent in their basic lifestyles if brave souls seek to purge themselves of it.

At the same time, this attempt to model purification in an impure world comes with the very serious danger of self-righteousness, as the ethicist Laura Hartman helpfully observes. She notes that in a world of climate change, it is a great virtue to take the bus or ride a bicycle rather than drive. However, it would be a failure to "attach great pride to the fact that I avoided using my car. . . . I could equally seek to gain attention by showing off the excellent new bicycle I purchased, riding it ostentatiously by my neighbor's house." She also learns from Woolman that there is no easy end to the moral questions. Riding a twenty-first-century bus, Woolman "might wonder about the treatment of the bus drivers, and whether paying the fare (which is fairly cheap) might contribute to their underpayment."[40] Anyone who seeks to follow in Woolman's footsteps should beware of the temptation to become self-righteous. We can fight against self-righteousness, however, if we notice how we fail to live up to our own ideals. Like Woolman, little should content us.

Looking back from two and a half centuries, one can identify failings even in Woolman's witness. For example, he was concerned about the violence inherent in a globalized economy, but his own travels and his own colonial life were utterly dependent upon that system.[41] With the benefit of hindsight, one might also raise questions about Woolman's approach to abolition, noticing that he focused on white slave owners and paid little attention to empowering and supporting slaves and former slaves.[42]

Woolman would almost certainly have listened carefully to such critiques. He knew that he was not perfect, that he remained complicit in the trappings of wealth and privilege. He did not believe he needed to be pure, only that he should be seeking purification. He writes that perfection is "a goal remote," and that "the place to which we trend is not within our view." He took comfort that "every step we take is an abatement of the distance," but he never claimed to have arrived at purity.[43] He did not believe that human beings are called to be blameless but rather that everyone is called to always be working toward a better life.

Woolman sought to live as well as he could in a broken world, and he contributed to an expanding awareness of the violence in that world's structures. Most people who live in the industrialized world will become paralyzed if they seek to completely avoid any complicity in the violence of climate change. As Baird Callicott points out, concerned people run the risk of being powerlessly silent if we refuse to use the tools of today's society to speak against its failings. As Martin Palmer points out, concerned people will have trouble recruiting others to our cause if we "forget how to party" while seeking to live out our principles. This is why Woolman's witness is so powerful; it gives a model for how people can separate themselves somewhat from the structural violence of their time while still accepting their moral finitude. We cannot become perfect, and we cannot solve the problem of climate change. But we are not powerless, and can do better.

### Empathy for Those Who Suffer

Meaningful opposition to climate change will require deep and profound empathy. Those who resist structural violence need the ability to understand and relate to others' experiences and feelings. Wealthy citizens of the industrialized world must learn to feel the urgency of other people's suffering, of other species' extinction, of degraded ecosystems from which our selfishness and society have increasingly separated us.

John Woolman is a model of empathy, and his commitment to purification and self-denial was in many ways a training program to help him better understand the suffering of others.[44] Interestingly, part of his argument against slavery and high concentrations of wealth was that both degraded empathy. Those who owned slaves or paid others to do all their labor lost the ability to understand hard work and with it the ability to empathize with hard workers.[45] As always, Woolman practiced what he preached; although he was a talented merchant and the son of a prosperous farmer, he sought material simplicity in order to better empathize with the poor.

In a culture that had learned to ignore the suffering of African slaves, Woolman asked his readers to imagine their own children and friends kidnapped and shipped across the ocean. In a time when the abolition of slavery seemed an impractical dream, he was already arguing that justice demanded not only an end to slavery but also repayment for the unpaid labor of the enslaved, if necessary paid to their descendants by society at large.[46]

Woolman's ability to relate to and empathize with the pain of others is a vital example for contemporary people who care about climate change. Those who live in wealthy areas protected from rising seas and droughts need to learn to empathize with Bangladeshi families losing land to encroaching saltwater and Sudanese communities increasingly threatened by war and famine in a drying climate. All human beings need to find ways to relate to, care about, and act to prevent the suffering of polar bears who are losing their hunting grounds. Like Woolman, concerned people today should examine our luxuries and ask whether they make it harder to understand others who suffer. Like Woolman, we should nurture an imaginative empathy so that we can understand those who suffer and thereby treat them more justly.

For example, Laura Hartman is inspired, partly by John Woolman, to argue that people in the twenty-first century should considering no longer eating meat, in part because of the "egregious environmental ills" associated with its production. Interestingly, however, one of the central hesitations Hartman admits about vegetarianism also emerges from empathy; to relate to others often means to eat the food to which they are accustomed. She writes, "it would certainly offend my host in a visit to Haiti if I refused to eat the goat that he and his family had graciously, and at great cost to themselves, prepared for me."[47] Empathy for the victims of climate change, and for goats, is here in tension with empathy for a culture that sees the sharing of goat meat as an act of hospitality. Here one faces moral finitude. In such a situation, one should seek not perfection but the best answer available in any particular circumstance.

However, given that meat usually costs more than its alternatives, eating less meat is likely a good way to empathize with the world's poor. Industrial meat production contributes to climate change; but, just as important, it distances human beings from other creatures. Eating less meat is a constructive response to climate change not only for its practical effects on the climate but also for the empathy it cultivates for those others, both human and animal, who suffer.

## Empathy for Perpetrators

Empathizing with those who suffer is only one part of the work ahead. Woolman worked to relate not only to the poor and the enslaved but also to the rich and to slave owners. He tried to give them the benefit of the doubt, and he imagined that they were motivated by the good intentions of seeking to build up wealth for their children, fuel economic prosperity, or even offer a more "civilized" life for slaves of African descent. Though he disagreed with these arguments, he could meet those who made them with respect; he could insist that even when he argued against them, he was not dismissively judging them. He observed that if he respected the reasoning of others with whom he disagreed, they were more likely to listen to his counterarguments.[48]

Woolman points out that if one seeks to find the truth, it is best to assume that even those with whom one disagrees might have a part of it. Concerned people have more reasons to openly share our ideas and influence with others when we consider the possibility that we ourselves could be wrong.[49]

Contemporary activists seeking to learn from Woolman should be challenged by his willingness to listen to and even respect the arguments of slave owners. With the benefit of historical distance, people today know beyond a doubt that slavery is horrific, and it may be difficult to imagine respectfully considering the logic of a person who claims the right to own, control, and even torture another human being. But perhaps this difficult notion is a particular gift from this witness; this may be the place at which Woolman can teach the most.

Michael Birkel explains Woolman's approach in very straightforward terms: "Because we are all so connected, and because we all fall short of our

ideals, it is unwise and untrue simply to blame others as wrong."[50] Woolman's empathy for the perpetrators as well as the victims of structural violence came from his awareness that he himself was a perpetrator, that even
one who worked as hard as he did could never fully live up to his own ideals. Thus, he managed to purify himself and create a powerful witness for
justice "without succumbing to the pitfall of self-righteousness."[51] He knew
that he was not perfect, and so he never took upon himself the responsibility
of judging others for their failings. He would argue with them and explain
himself carefully, but he never asserted a certainty of his own righteousness.
Instead, he focused on setting the best example he could, trusting the good
within them to recognize and learn from the good within himself.

Woolman's awareness of his own moral finitude has much to teach
those of us who are concerned with climate change. Our empathy for
those who suffer from climate change must fuel our commitment to make
changes in our own lives, in our political structures, and in today's world.
But Woolman's example suggests that empathy would be undone by self-
righteousness, and his witness suggests that climate activists might also
need to learn empathy for oil executives, advocates of corporate agriculture, and private jet owners. There is much to argue about with such people, and we should seek to change their minds. But we cannot dismiss
their perspectives and arguments because we, too, are imperfect. We, too,
are changing the atmosphere. We share profound guilt for the structural
violence of climate change.

In 2012 the journalist and activist Bill McKibben published an important essay, "Global Warming's Terrifying New Math," in which he observed
that the only way to keep the global temperature from rising more than
2°C would be to leave most remaining oil, natural gas, and coal in the
ground, despite the fact that fossil fuel companies were already counting
these materials as assets. He suggested that these facts reveal "who the real
enemy is" in the fight against climate change. For him, the fossil fuel industry is "Public Enemy Number One to the survival of our planetary civilization," and he sees this as good news for the climate movement, which
will be more effective at organizing with a clear force to organize against.[52]

John Woolman's example raises a caution here. Climate activists may
productively view the fossil fuel industry as an enemy, but only if they
can also cultivate empathy for the human beings in that industry. Most of
them are motivated not by an evil desire to destroy the planet but instead

by reasonable—if limited—concerns for their own well-being, for their families, and for existing economic systems.

McKibben may be right about the climate movement's strategic need for "an enemy," but seeing other people as enemies is, at best, a partial path toward climate justice. Recall that Gandhi believed that it was better for people to strike out against injustice in anger and hatred than to quietly acquiesce to it. If the choice for the climate movement is to either have an enemy or accomplish nothing, then an enemy is the better choice. But Woolman shows that there is a higher goal, a nonviolent ideal whereby one resists the violence of climate change without resisting any other human being, without dismissing any other person. This is consistent with Bill McKibben's own activist work, as demonstrated by his role in many important nonviolent protests such as the 2014 arrest action in opposition to the Keystone XL pipeline. Two years after he called for an enemy, McKibben also insisted that nonviolence is "the most important technology" developed in recent history, and insisted "that we will fight, in every corner of the earth and with all the nonviolent tools at our disposal. . . . We will see if that new technology of the twentieth century will serve to solve the greatest dilemma of our new millennium."[53]

Woolman suggests that a movement ultimately needs empathy even more than it needs enemies. True and abiding change is much more likely if those who extract fossil fuels have some reason to listen to those who want them to stop. Dismissing the enemy is a good way to stop them from listening to what I have to say. The witness of Woolman suggests that care for all creatures should lead people to genuine empathy, even for those who are harming the climate. More empathy will lead to a stronger example, to more pervasive change. Climate activists should take this witness seriously in the work ahead.

## LIVING MORALLY IN A FINITE WORLD

Inspired by John Woolman, the Quaker environmentalists Louis Cox and Ruah Swennerfelt spent six months walking across the West Coast, from Vancouver to San Diego. They stopped along the way to speak with Quaker meetings about care for the Earth. The ideas they gathered for and from that journey are powerful resources for anyone who seeks inspiration and guidance from Woolman about climate justice. Cox and Swennerfelt write that he set an example with "radical changes in his own life once he realized that wealth and luxury are idols that distract us from that holy calling, building the Peaceable Kingdom."[54]

Woolman believed that all people are called to moral purification and that they have within them the impulses and willpower required to pursue that purification. And yet he also saw that human greed and structural violence can blind people to this capacity for peace within themselves. He responded by offering a public witness against slavery and every other injustice he saw, not only resisting the trappings of structural violence but also ensuring that others would notice his calling and, hopefully, work harder to pursue their own.

Cox and Swennerfelt pursued their callings by walking for six months to emphasize the continued relevance of Woolman's work in a time of changing climate. John Francis and Eric Holthaus pursued their callings by serving as examples of the possibility of living without transportation based on fossil fuels. These are powerful witnesses that have much to teach anyone resisting climate change.

Ultimately, I find Baird Callicott and Martin Palmer, the critics of personal purification cited earlier in this chapter, less convincing. I disagree with them that the key danger facing climate activists is disengagement from the culture or an appearance of dullness. Nevertheless, I have worked to learn from and listen empathetically to their arguments. Both work hard to understand what is happening to the climate and to spread awareness of the problem and its causes, which I appreciate.

I still worry about their arguments that personal sacrifice is not a helpful part of such a witness. Callicott proudly decries his previous efforts to become "Joe Bioregionalist," and Palmer dismisses the efforts of a couple who refuse to fly in planes as those who "forgot how to party." Learning from Woolman's witness helps me to recognize the limitations of these perspectives. In my view, Callicott and Palmer do not take seriously enough what can be taught by a sincere effort at purification. By deemphasizing the personal dimensions of a moral response to climate change, they remove an important tool for genuine resistance. All privileged people who care about climate change can demonstrate our commitments by purifying ourselves.

One path to purification already discussed is eating more locally and lower on the food chain. Doing so, people eat more efficiently, their food is produced with less energy, and it costs less energy to get food to them. They also provide less support to the industrial agricultural systems that change the climate. I am far from perfect, but I have had some success in this category by removing most meat and dairy products from my diet. I hope that I have been able to reduce my contribution to climate change slightly by making responsible choices and by modeling them to others while doing my best to avoid self-righteousness. However, moral finitude makes this challenging in many ways, and many people with different health conditions, from different cultures, or who live in different climates will have a harder time making such changes.

Another important step would be to forgo or limit transportation by methods that contribute inordinately to climate change. This is a powerful witness and deserves careful consideration. I confess that I have had little success here. I sometimes ride the bus, and I carpool when possible, but these very minor choices are more than counterbalanced by the half dozen plane trips I take each year. Some trips are for the same reason as Callicott's; like him, I learn and teach about climate change in a profession that assumes mobility, and those with whom I share ideas live all over the nation and the world. Thus, I travel in part to spread the word about the damage I am causing by my travel. However, other trips I take are to see my biological and chosen families. I am deeply committed to people who live far away from me, and I am currently unwilling to stop enjoying their company in person. This is a point at which I face my moral finitude; the world as it currently exists gives me no easy way to maintain my relationships while avoiding the harm of fossil-fueled travel.[55] I continue working to do better.

John Woolman has inspired another attempt to change my own life, as I try to wean myself from air conditioning.[56] Using fossil fuels to cool the air in my home not only releases $CO_2$ into the atmosphere but also increases the heat around my home, adding to an already hot environment. In large cities, this creates a cascading "heat island" effect that consumes yet more energy and makes people more uncomfortable. These are good reasons to reduce my use of air conditioning. However, Woolman points to another, perhaps more important, motivation: empathy. The more I shelter myself from the weather, and especially from increasingly hot summers, the less immediate my own experience of climate change will be. I will be slightly more able to empathize with those suffering heat waves across the world if I do not use fossil fuels to unnecessarily change the temperature in my immediate vicinity. Thus, I seek to reduce my use of air conditioning as one small step toward purification from the violence of climate change.

One of John Woolman's most powerful forms of activism came in the conversations he had with others. When he shared his discomfort about writing a will that treated a human being as property, the slave owner who had asked him to write the will had an opportunity to talk about his own discomfort. When Woolman shared his concerns about luxury with rich acquaintances, he offered them a chance to follow their better impulses. Whatever other changes we make, we can all follow Woolman's example by finding ways to talk with those around us about the structural violence of climate change. We can share the ways we worry about our daily habits. To be strictly vegan reduces one's impact considerably. But perhaps being willing to bend those rules for good reasons now and then is worthwhile if, when doing so, one gently shares concerns and allows others to reflect on their own habits.

Few concerned people will be able to do this all the time. I have yet to start a conversation about the violence inherent in contemporary travel with a stranger on an airplane. But I am trying to talk with those close to me about my concerns. All people can share these moral struggles some of the time.

At the root of moral problems, John Woolman believed, is a selfish refusal to listen to the best impulses within oneself. This selfishness lures many of us to continue enjoying the comforts of a middle-class lifestyle in the industrialized world, eating whatever we want, flying regularly on planes, running an air conditioner when it gets hot, and politely keeping our opinions to ourselves so as not to be rude. But the better and purer voice within us calls for a different way of life, a resistance to cultural patterns that cause violence. If we listen to the best voices within ourselves, this will help us to cultivate empathy for those who suffer as the climate changes, and will help us to set an example of resistance against the systems that are responsible.

Sharing one's concerns, changing one's diet, and making more responsible transportation choices will not end climate change. In part, this means that people must also turn to broader institutional and social changes, which are the subject of subsequent chapters. Here, however, it is important to assert that personal change in light of climate change is nevertheless significant and called for. To resist the structural violence of climate change, concerned people must purify our own lives.

Everyone is finite, and no one can do it all. Honestly facing this fact is essential to a life of integrity in a world of climate change. If concerned people truly learn from Woolman's quest for purification, we will not judge one another or ourselves too harshly for our limits. His life was not about judgment but about truth. Thus, after facing our limits honestly, we can trust that we will have the power to do a great deal—to move toward purification.

## NOTES

For biographies of Woolman, see especially Slaughter, *Beautiful Soul of John Woolman*; and Plank, *John Woolman's Path*. For moral and theological analyses of Woolman, see Phillips Moulton, "John Woolman's Approach"; Heller, *Tendering Presence*; and Birkel, *Near Sympathy*.

1. Woolman, *Journal and Major Essays*, 33.
2. Ibid, 236.
3. Ibid., 24.
4. Ibid., 24–25.
5. Ibid., 35.
6. Birkel, *Mysticism and Activism*, 15.
7. Woolman, *Journal and Major Essays*, 200.
8. "He who with view to self-interest buys a slave made so by violence, and only on the strength of such purchase holds him a slave, thereby joins hands with those who committed that violence and in the nature of things becomes chargeable with the guilt." Ibid., 233.
9. "Harken then, Oh ye children who have known the Light, and come forth! Leave everything which our Lord Jesus Christ does not own." Ibid., 255.
10. As the historians Brycchan Carey and Geoffrey Plank note, Woolman and other abolitionists of his time can be critiqued for not entirely living out this last idea because his primary conversation about slavery was with white slaveholders rather than with slaves themselves: "Enslaved men and women made their unhappiness clear, but Quaker masters began to free them only in response to pleas from other white Quakers. The Quaker antislavery movement began as a conversation among whites." Carey and Plank, *Quakers and Abolition*, 1. Jean Soderlund offers an even harsher critique of the approach Woolman helped to shape: "The gradualist, segregationist, and paternalistic approach of Friends set the tone for the white antislavery movement in America from 1780 to 1833." Soderlund, *Quakers & Slavery*, 185.

11. As the historian Geoffrey Plank puts it, "Woolman understood the power of saintliness" and so offered a "detailed and sweeping critique of the material culture and economy of the British empire" through his own life. Plank, *John Woolman's Path*, 3.

12. Trueblood, *People Called Quakers*, 162. This story is, as far as I know, only told by Trueblood. Perhaps it did not occur, or Trueblood has embellished it. However, it effectively captures the style of quiet moral protest for which Woolman remains legendary.

13. See Woolman, *Journal and Major Essays*, 119–22.

14. This process of purification was a fundamental practice of the Society of Friends in which Woolman was raised. See Birkel, *Mysticism and Activism*, 8–14.

15. Woolman, *Journal and Major Essays*, 236.

16. Ibid., 59. Along similar lines, he helped to write an epistle from the Pennsylvania and New Jersey General Meeting that called on all Quakers "to have our minds sufficiently disentangled from the surfeiting cares of this life, and redeemed from the love of the world, that no earthy possessions nor enjoyments may bias our judgments, or turn us from that resignation and entire trust in God, to which his blessing is most surely annexed." Woolman, *Affairs of Truth*, 18.

17. Ibid., 156.

18. Woolman, *Journal and Major Essays*, 202–3.

19. Ibid., 205–6.

20. Ibid., 212. Woolman connected the issue of slavery to the issue of poverty, noting the same dynamic: as children of the rich were taught that they deserved excess, they were also taught to ignore the ways that "the faces of the poor have been ground away, and made thin through hard dealing." Woolman, *Affairs of Truth*, 151.

21. Woolman, *Journal and Major Essays*, 57. The idea of becoming a "fool" is a reference to 1 Corinthians, which says that "the wisdom of this world is foolishness with God" (3:19) and "We are fools for the sake of Christ" (4:10). This was also a favorite biblical reference of Dorothy Day, to be discussed in chapter 5.

22. Ibid., 204–5.

23. "Al Gore's Personal Energy Use Is His Own 'Inconvenient Truth,'" February 25, 2007, www.beacontn.org/2007/02/al-gore%E2%80%99s-personal-energy-use-is-his-own-%E2%80%9Cinconvenient-truth/.

24. Henning, *Riders in the Storm*, 156, 159. Henning does not use Gore as an example and so does not answer the question of whether the former vice president is best understood as hypocritical or morally finite. For two defenses of Gore, see A. C. Grayling, "Good Men in a Bad, Bad World," *The Independent*, March 4, 2007; and Aiken, "Gore's Purported Hypocrisy."

25. Eric Holthaus, "Why I'm Never Flying Again," 2013, http://qz.com/129477/why-im-never-flying-again/.

26. Francis, *Planetwalker*, 7–8.

27. Ibid., 218.

28. Callicott, *Beyond the Land Ethic*, 46–47.

29. Ibid., 51.

30. The quotation from Palmer and the interpretation are from Johnston, *Religion and Sustainability*, 146–47.

31. Moulton, "John Woolman's Approach," 408. Moulton also takes pains to emphasize that Woolman was not merely purifying himself for strategic reasons, emphasizing that Woolman believed "personal purity was of value for its own sake" (p. 403), and that this commitment to purification as an intrinsic good was in fact part of what gave him such obvious integrity in the eyes of others.

32. McFague, *Blessed Are the Consumers*, 42.

33. Bales, *Disposable People*.

34. See especially Davis, "Woolman and Structural Violence," 243–60.

35. Woolman, *Journal and Major Essays*, 114.

36. Orr, *Nature of Design*, 144, 147. Orr takes the analogy farther than I would, suggesting, for example, that "our dereliction will be judged a more egregious moral lapse than that which we now attribute to slave owners." Though I agree with him that wealthy people in the industrialized world are guilty of "knowingly bequeath[ing] the risks of global destabilization to all subsequent generations everywhere" (p. 146), I worry that such rhetoric becomes too dismissive of difficult arguments to be had about how to respond to climate change, tempting one to the simplistic righteousness that Woolman argued against.

37. Birkel, *Near Sympathy*, 61.

38. Plank, *John Woolman's Path*, 200.

39. Woolman's biographer, Thomas Slaughter, writes that Woolman's "quest for purity drove him to both harmless eccentricity and self-destructive extremes." Slaughter, *Beautiful Soul of John Woolman*, 10. The philosopher Phillips Moulton agrees that Woolman "developed a few idiosyncrasies that sometimes appear more like defects than virtues," and he admits that Woolman's "unswerving integrity" could appear naive, that "perhaps he was overscrupulous on occasion." "Introduction," in *Journal and Major Essays*, by Woolman, 16.

40. Hartman, *Christian Consumer*, 186.

41. Both these issues are discussed at length by Plank, *John Woolman's Path*, esp. 32, 297.

42. Although I do not know of anyone leveling this critique at Woolman specifically, Jean Soderlund develops the argument about Quaker abolitionists in general. See especially Soderlund, *Quakers & Slavery*; and Nash and Soderlund, *Freedom by Degrees*.

43. Woolman, *Affairs of Truth*, 102.

44. Philip Boroughs writes that one can see in the *Journal* a "gradual movement from sympathy to empathy." Concerned for the poor and oppressed, Woolman increasingly changed his own life to share their reality and became more sensitive to not only the needs but the full realities of others. Boroughs, "John Woolman's Spirituality." in *Tendering Presence*, ed. Heller, 13–14.

45. Woolman, *Journal and Major Essays*, 241–42.

46. Indeed, Woolman seems to advocate what we would today call reparations for slavery. Consider: "Having thus far spoken of the Negroes as equally entitled to the benefit of their labour with us, I feel it on my mind to mention that debt which is due to many Negroes of the present age. Where men within certain limits are so formed into a society as to become like a large body consisting of many members, have whatever injuries are done to others not of this society by members of this society, if the society in whose power it is doth not use all reasonable endeavours to execute justice and judgment, nor publicly disown those unrighteous proceedings, the iniquities of individuals becomes chargeable on such civil society to which they remain united." Ibid., 270.

47. Hartman, *Christian Consumer*, 189–90.

48. Woolman, *Journal and Major Essays*, 211. Woolman makes similar comments about fellow Quakers who, unlike him, were willing to pay taxes that support the military. Woolman insisted that he could both live out his principles and respect others who thought differently: "I all along believed that there were some upright-hearted men who paid such taxes, but could not see that their example was a sufficient reason for me to do so, while I believed that the spirit of Truth required of me as an individual to suffer patiently the distress of goods rather than pay actively" (p. 75).

49. For further discussion of Woolman's "politics of empathy," see Myles, "'Stranger Friend,'" 45–66.

50. Birkel, *Near Sympathy*, 9.

51. Ibid., 97, 98.

52. Bill McKibben, "Global Warming's Terrifying New Math," *Rolling Stone*, August 2, 2012. Four years later, McKibben changed his articulation of the "enemy" somewhat when he wrote that the struggle against climate change is "a world war" fought against carbon. See McKibben, "World at War." For an argument against martial rhetoric in the struggle against climate change, see O'Brien, "'War' Against Climate Change."

53. Bill McKibben, "How the Active Many Can Overcome the Ruthless Few," *The Nation*, December 19–26, 2016.

54. Cox and Swennerfelt, *Walking in the Light*, 8.

55. George Monbiot refers to this as the problem of "love miles: the distance between your home and that of the people you love or the people they love" and suggests

that this is a primary reason for widespread denial about the climate effects of airplane travel. Though the term is helpful, Monbiot does not intend to excuse travel for purpose of community: "It has become plain to me that long-distance travel, high speed, and the curtailment of climate change are not compatible. If you fly, you destroy other people's lives." Monbiot, *Heat*, 172, 188.

56. This move was also inspired by Pope Francis's 2015 encyclical on the environment, which calls out the use of air conditioning as "self-destructive" behavior. Pope Francis, *Laudato Si'*, §55.

# 4

## Jane Addams and the Scales of Democracy

We are bound to move forward or retrograde
together. None of us can stand aside; our
feet are mired in the same soil, and our lungs
breathe the same air.
    —Jane Addams, *Democracy and Social Ethics*

In April 1895, at the age of thirty-four years, Jane Addams took the first paid job of her life: garbage inspector for Chicago's Nineteenth Ward. The garbage collectors in her neighborhood, who were appointed through political favors, routinely neglected their work. The streets were so over-run that children were playing in trash and dead animals were left rotting on the side of the road. Addams hired an assistant, and the two of them woke at 5 am to follow the collectors three mornings a week, ensuring that the garbage was picked up on schedule. They also advocated for expanded service, set up new incinerators, created a recycling program for tin cans, and arranged a process for the removal of dead horses and cows.[1]

Twenty years later, Addams helped to found and became the first president of the Women's International League for Peace and Freedom, and in this capacity she developed principles for global peace, advocated the creation of international organizations to prevent war, and advised

President Woodrow Wilson on the peace treaty ending World War I. She was awarded the Nobel Peace Prize in 1931. Yet she remained a resident of Chicago's Nineteenth Ward, and she always stayed connected to her neighborhood and community.

Addams's cause, locally and globally, was democracy. To her, democracy meant that every person matters and all people are connected to one another. This was not just a political idea, but also had implications for economics, culture, and family—in all spheres of life, she argued, every voice should be heard and valued. As she wrote in her first book, *Democracy and Social Ethics*, "Our feet are mired in the same soil, and our lungs breathe the same air." Accepting this common ground, she wrote, "brings a certain life-giving power," because when people understand "that we belong to the whole," and that all others do as well, then "a certain basic well being can never be taken away from us whatever the turn of fortune."[2]

Those of us concerned about the violence of climate change have much to learn from Jane Addams's commitment to democracy, and perhaps the most important lesson is the way she advocated for it across diverse scales. The multicultural community of her neighborhood helped her to better imagine planetary peace, and negotiating international agreements helped her to better understand the lives of her immigrant neighbors. To resist the violence of climate change, concerned people need to think and work across local and global problems and consider every scale in between. We need to use the personal changes discussed in the previous chapter to not only influence our neighbors but also to shape international politics. We need to learn from national movements about local issues. Climate change is a problem around the planet and in particular neighborhoods, so we need to learn to advocate democratic solutions on every scale.

## LEARNING TO BE A GOOD NEIGHBOR

Jane Addams was born into a prominent and prosperous family in rural Illinois in 1860. Her father was a successful miller and a state senator who was friendly with Abraham Lincoln. Jane attended Rockford Female Seminary, where she was a star student, and then she spent a few years considering medical school, traveling in Europe, and continuing her studies. She describes these years after college as deeply frustrating because she was well educated but could find no productive way to use her education. She realized with disappointment that her society taught wealthy people to indulge in leisure rather than work and encouraged women to improve their homes but not the wider world.

Addams resisted both of these cultural norms, and she describes two experiences in Europe that changed the course of her life. The first was a bullfight in Madrid, where she watched "with comparative indifference" as five bulls and many horses were killed. When the women with whom she was traveling pointed out that she was oddly unaffected, she was disturbed by her own complacency. She writes, "I felt myself tried and condemned, not only by this disgusting experience but by the entire moral situation which it revealed." She worried that her culture and her schooling had taught her to be a passive observer, to see violence and oppression in the world but not to be personally unsettled by them or to take any action to stop them. But now she made a commitment to change—to learn how to feel more deeply and act more resolutely.

Her second experience was a visit to Toynbee Hall, a famous "settlement house" where wealthy British men lived in an impoverished London neighborhood in order to learn about the poor and help develop solutions to their problems. She began to believe that such a lifestyle would resolve her listlessness and moral blindness. To live with the poor would give her "the solace of daily activity."[3]

In 1889, Addams pooled her resources with those of a college friend, Ellen Gates Starr, to found Hull House, a settlement on the Near West Side of Chicago. This was a neighborhood increasingly populated by European immigrants, and in a time of expanding industrialization and a volatile economy, many of these immigrants struggled to find work and make ends meet. Starr, Addams, and other educated women from wealthy families lived at this settlement and volunteered their time to run programs and offer courses for their neighbors. Men joined as residents a few years later.

Hull House was motivated by religious impulses but was not a Christian institution. Addams grew up in a home shaped by Christianity, but her father was never dogmatic about religion. Though she attended a Christian college, she bristled at much of the theology taught there. However, shortly before moving to Chicago, she was baptized as a Presbyterian, and she frequently wrote about how the settlement exemplified the Christian ideal of love for neighbors. Her cofounder, Ellen Starr, was far more devout but nevertheless agreed with Addams that Hull House should be inclusive and welcoming of diverse neighbors, including those who were not Christians. They decided that, unlike Toynbee Hall and other prominent settlements, Hull House should not be an officially Christian project.[4]

In her earliest explanations of Hull House, Addams distinguished between the "subjective" and "objective" needs that it met. The struggling residents of Chicago's poor neighborhoods had objective, external needs. They needed education, day care, and political voice. The wealthy, by contrast, had subjective, internal needs. They needed to find purpose and discover real-world applications for their educations. Hull House was, therefore, "an attempt to relieve, at the same time, the over-accumulation at one end of society and the destitution at the other." Addams understood the rich and the poor as inextricably linked, in need of one another. Writing to privileged people like herself, she insists that a democratic society can only thrive if everyone does well: "The good we secure for ourselves is precarious and uncertain, is floating in mid-air, until it is secured for all of us and incorporated into common life."[5]

Hull House, as Addams understood it, was primarily an educational institution, comparable to a university or college and dedicated to bringing scholarship and culture to people who would otherwise be left out of "the circle of knowledge and fuller life."[6] Therefore, she and her colleagues organized book clubs, lectures by local academics, and discussion groups. They discovered that many working people, even those exhausted by long hours of factory work, were hungry for intellectual engagement. The residents also created a kindergarten and day care program, educating and empowering thousands of children. After forty years of this work, Addams wrote of a proud moment when a journalist told her that Hull House was "the first house I had ever been in where books and magazines just lay around as if there were plenty of them in the world. . . . To have people regard reading as a reasonable occupation changed the whole aspect of life to me, and I began to have confidence in what I could do."[7]

After years of this work, Addams identified flaws in her original understanding of settlement life. In hindsight, she realized that her early motivations for Hull House had been patronizing. She had assumed that her neighbors would be passive recipients of aid, gratefully receiving what the educated settlement residents had to offer. But she quickly learned that poor people, just like rich people, had subjective needs and sought purpose and engagement with their neighbors. Although she was privileged, she had much to learn from her neighbors. This became a key moral principle animating her activism: Anyone who seeks to help others must learn from them, working with rather than on behalf of those in need.

Aware of her own limitations, Addams made a rule to never speak about Hull House's work without inviting a neighbor to join her, "that I might curb any hasty generalization by the consciousness that I had an auditor who knew the conditions more intimately than I could hope to do."[8] Thus, Hull House's neighbors became not only recipients of programs but also collaborators, teachers, and friends.

In conversation with their neighbors, the residents of Hull House quickly saw the need for programs beyond their educational mission, and they began offering meeting spaces for community groups, assistance with legal and political struggles, and advocacy on issues like the neighborhood's mounting trash that Addams helped to solve when she served a year as a garbage inspector. Her prominence and fame spread, and she was soon a respected figure in Chicago politics and society. This influence continued to expand, and by the turn of the century she was an internationally sought-after author and speaker. She became a valued conversation partner for notable intellectuals like the pragmatist philosophers John Dewey and William James.

Jane Addams lived in a society that encouraged women to demurely focus their attention on family and home life. Whether she privately agreed with such treatment of her sex is not known, but in public she did not challenge these norms. Instead, she used society's expectations that women should focus on the private life to justify her very public efforts. She argued that her settlement work, her political activism, and her international advocacy were all logical extensions of feminine tasks. For example, when neighbors protested that garbage inspection was not a role for women, she insisted that she was simply doing the woman's work of helping families stay clean and healthy. When she began to advocate for women's suffrage, she insisted that women deserved a voice not for their own sake but so that they could advocate for children at the ballot box. She cleverly characterized such extensions of women's role into the public sphere as "civic housekeeping." Addams, who had no biological children, thereby took on the "motherly" role of household manager and caretaker to justify her activism.[9]

The same thinking fueled her peace activism. As she saw it, wars threatened not only the lives of the men fighting them but also the harmony of families and households. Thus, it was appropriately woman's work to negotiate the prevention of wars and to reform politics so that all voices could be heard and world peace could be more possible. In 1906 she published *Newer Ideals of Peace*, which argued that simplistic, "dove-like" views of peace, such as the simple absence of conflict, were inadequate. Instead, she drew on her experiences to advocate a "more aggressive," "active and dynamic" view of peace that prevented conflict by empowering people, ensuring that their basic needs were met, and building just governing structures.[10]

As World War I began, Addams helped to found the Women's International League for Peace and Freedom. After the war, she continued to devote her attention to this cause, and she insisted that peace could not be separated from other struggles for justice. She argued that wars emerge from poverty and desperation, and so the best way to prevent them is to ensure that all people are educated and fed.

She also supported international laws outlawing war. She knew that this idea sounded naive to many, but she insisted that if human societies had learned that slavery was fundamentally abhorrent, they could learn the same about war. Offering news that would have cheered John Woolman, she wrote that "slavery has joined cannibalism, human sacrifice, and other once sacred human habits, as one of the shameful and happily abandoned institutions of the past."[11] She believed that hard work could ensure that future generations would say the same about war.

Addams still resided in Chicago and was still involved in local social activism and the struggle for international peace when she died at the age of seventy-three in 1935.

## THE LEVELS OF LIVED DEMOCRACY

One cannot understand the life and work of Jane Addams without attending to the idea that animated her: democracy. Her friend and colleague John Dewey credited her with the idea that democracy is not just a political system but also a complete "way of life" that places social, cultural, and economic as well as political authority into the hands of every person.[12]

Democracy, as Jane Addams understood and practiced it, is about living as though people can take care of themselves when they are trusted to do so.

This principle motivated her to advocate for global peace and to move to Chicago's Ninth Ward. She believed that all people should have the right to live fully into their potential, participate in the decisions that shaped their lives, and benefit from community along with their neighbors and the rest of humanity. The structural violence of poverty in Chicago's neighborhoods prevented the children of immigrants from living up to their potential, and so Addams resisted poverty. The direct violence of war around the world prevented soldiers and their families from entering into community with others, and so she resisted war. The world and her neighborhood were violent and unfair, and so she devoted her life to bringing democratic equality to both.

Addams even explained her Christian faith as an expression of her "passionate devotion to the ideals of democracy." She celebrated the theological idea that everyone is equal before God and the resulting faith's resistance against unnecessary hierarchies and domination. Reflecting on the poverty of Jesus's own background, she asked, "when in all history had these ideals been so thrillingly expressed as when the faith of the fisherman and the slave had been boldly opposed to the accepted moral belief that the well-being of a privileged few might justly be built upon the ignorance and sacrifice of the many?"[13]

Addams also described the foundational motivation of Hull House as an "attempt to socialize democracy," to ensure that everyone was fully engaged in and empowered by her community. The founding ideal of the settlement house was to offer poor people resources to engage in culture and politics while providing privileged people with a way to get to know their neighbors.[14] For this reason, she resisted anyone labeling Hull House as a charitable or philanthropic institution. She worried about "an unconscious division of the world into the philanthropist and those to be helped" and instead insisted that a truly democratic effort would ultimately be for the sake of all people, recognizing their connections and their capacity to learn from their differences.[15]

This kind of democracy depends upon expanding moral attention, moving from individual concerns about personal virtue and family well-being to what Addams called "social morality," which includes concern for the common good and a belief that all people depend upon and can learn from one another. This means not only caring about others but also seeking to understand them on their own terms. In *Democracy and Social Ethics*, Addams argues that democracy cannot be enacted by anyone alone; it requires an education about and from other people, a deep acquaintance with "that diversified human experience and resultant sympathy which are the foundation and guarantee of Democracy." When people become acquainted with this diverse experience, she believed, they will come to understand that individual needs and individual morality are utterly dependent upon communities. Individual morality is "widened until it gradually embraces all the members of the community and rises into a sense of the common weal."[16]

Addams learned this lesson about "widening" moral concern as economic and technological changes were threatening to narrow the experience of workers in the industrialized world. In factories, each person's tasks were becoming more and more specialized. This division of labor made people more efficient and more interdependent, but she worried that it could also "rob" them "of any interest" in their work: "The man in the factory, as well as the man with the hoe, has a grievance beyond being overworked and disinherited, in that he does not know what it is all about."[17] Thus, Addams observed, just as she and her peers had needed to find a way to use their education in service to others, so too did her neighbors need to understand the significance of their labor. This did not mean that a higher purpose could justify mistreating one's workers—Addams was adamant in supporting the basic rights of all people. But she insisted that, ultimately, individual rights are never enough. Democracy only works when everyone feels like a part of their community rather than an individual separate from it.

Her peace activism was based on the same democratic awareness of interconnectedness. During the Spanish-American War, she noticed that her previously peaceful neighborhood had seen seven murders in just two months, which she credited to the "carnage and bloodshed" reported from the war. As she watched, "little children on the street played at war and at killing Spaniards," degrading their awareness of common humanity in ways that she feared would hurt them and their communities for years to come.[18] War is undemocratic, according to Addams, because it destroys communities and impedes the moral development of everyone within them. This was an observation about both a distant war and her own neighborhood.

Addams's democratic attention repeatedly used personal experience to expand moral concern. For example, she only realized that the garbage in her neighborhood posed a health threat when her nephew came to live with her and took ill. She sent her nephew away while she dealt with the problem but was "ashamed that other delicate children who were torn from their families, not into boarding school but into eternity, had not long before driven me to effective action."[19] An expert at admitting to and learning from her mistakes, she worked harder to pay attention to the needs of her neighbors, to think beyond her own family, and so to behave more morally, more democratically.

However, Addams did not merely want to scale moral attention upward. She also believed that the local and personal have much to teach about the global and the political. Arguing against World War I in 1915, she asked an audience in Carnegie Hall to learn from "any peasant woman who found two children fighting," who would say, "That can't go on; that leads to nothing but continued hatred and quarreling."[20] She also used Hull House's neighborhood as a model for international peace. If German and Russian immigrants could live cooperatively in Chicago, she argued, then Russia and the United States could find a way to make peace with Germany. Local examples of cooperation and peace between people of different faiths, different origins, and different social classes were models for global cooperation and peace.

Addams's pragmatic, multiscalar approach to democracy offers a distinction between her moral vision and John Woolman's. Although the eighteenth-century Quaker would almost certainly have celebrated her ideals, he responded to the violence of his time in a very different way, focusing on purification in order to model a better life for those around him. By contrast, Addams paid more attention to the institutions and structures that shape individual lives, focusing as much on national and international issues as on personal relationships. Though she learned to care deeply for her neighborhood, she never gave up the privilege of leaving it for lengthy vacations, and she remained wealthy all her life.[21]

However, Addams did consider living a life more dedicated to purification when she met one of her heroes, Leo Tolstoy, a member of the Russian nobility who famously lived as a peasant in order to follow Jesus. Tolstoy asked Addams to become more radical by joining in the manual labor of her neighbors. In response, she committed to spending two hours each morning in the kitchen cooking for the community. But she quickly gave up this idea, writing that "the whole scheme seemed to me as utterly preposterous as it doubtless was. The half dozen people invariably waiting to see me after breakfast, the piles of letters to be opened and answered, the demand of actual and pressing human wants—were these all to be pushed aside and asked to wait while I saved my soul by two hours' work at baking bread?"[22]

Addams and Woolman represent different responses to structural violence. Woolman, like Tolstoy, sought to purify himself to model a peaceful life. Addams sought instead to build democratic institutions that would nurture peace on a broader scale. However, the power of her witness is that though she worked more institutionally than Woolman, she remained a humble resident of Hull House who never dismissed the importance of individual action and personal choices. Her commitment to democracy integrates the personal, the local, the regional, the national, and the global. This is precisely the kind of activism required to develop a meaningful approach to climate justice.

## CLIMATE CHANGE AS AN ISSUE OF SCALE

Climate change is a multiscalar issue. Its violence is local, regional, national, and global. This creates a choice: Where should concerned people focus our attention and efforts?

Consider the village of Kivalina, the tip of a barrier island called home by more than 300 people in the far northwest of Alaska. The island has been a traditional site for trading and hunting among the native Iñupiat people for millennia, but the village was created by the Bureau of Indian Affairs in 1905 when it built a school and required previously nomadic natives from around the region to settle and form a town. The violence of that forced settlement is compounded today because the people of Kivalina are once again threatened. Their island is eroding as the sea ice that traditionally shielded it from ocean surges melts. Current projections suggest that, because of continued climate change, the island will be uninhabitable by 2025.

In 1992 the residents of Kivalina saw this problem coming and voted to relocate their village. But this relocation would cost tens or hundreds of millions of dollars, and they have spent the years since 1992 trying to raise this money. They requested the help of the US government but have not yet received any. In 2008 the village filed suit against twenty-four of the world's largest oil companies, arguing that their move is only necessary because these companies made massive fossil fuel extraction and burning possible and that the problem was exacerbated by these companies' campaigns of misinformation about climate change. Kivalina lost that case, with a judge determining that the blame for climate change cannot be pinned on any particular corporation and denying that the native peoples of Kivalina have the standing to even bring the case. At the time of this writing, the people of Kivalina continue to seek funding as the ice around their island melts and storms grow more severe.[23]

For the people of Kivalina, climate change is an immediate, local problem. But of course the ice around their island is shrinking because of global phenomena—an atmosphere changed by fossil fuels burnt and forests cut all over the world. The solution to Kivalina's problem is not solely local—its residents can only move if they are provided with resources from beyond their community. The most plausible solutions for Kivalina come from the national level because the US government that created the settlement could fund its relocation. But as I write, such national action seems unlikely, given current politics. International pressure, such as that exerted by a meaningful and enforceable global treaty on climate justice, could make national action more possible.

For a person concerned about climate change, the plight of Kivalina demonstrates the challenging moral questions of scale. As a perpetrator of climate change, do I as an individual owe the soon-to-be-displaced residents of Kivalina a debt? As a citizen of the United States, can I meaningfully advocate for them in the courts or in Congress? As a fellow human being, can I help them to be heard in the international community? At which of these scales should I devote my efforts?

These are complex and challenging questions. This section explores three scalar approaches to climate change broadly, and the following section shows how Jane Addams might help climate change advocates to work and choose between them. Although the focus of this chapter is on the academic question of moral scale, its conclusion will return to a more concrete reflection on how—and if—privileged people elsewhere can do something for the people of Kivalina, Alaska.[24]

## Thinking Bigger: Climate Change
## as a Global Problem

Most moral discussions of climate change assume that a constructive response to the problem will require scaling up, encompassing a larger "we" when considering how people live together. The logic is this: Previous generations may have survived with primary concern for their close kin, for their own neighbors, or for their community or nation. But now that industrialized peoples are changing the climate, only a moral community as large as the atmosphere itself makes sense. These arguments tend to emphasize the need to expand people's moral thinking, to enable them to think bigger, so that they can care more about other people than they have become accustomed to.[25]

For example, Pope Francis's 2015 encyclical, *Laudato Si'*, is addressed to "every person living on this planet" and calls for "one world with a common plan" in response to the integrated violence of climate change and poverty. The encyclical acknowledges that some decisions must be made at the local and national levels, and it mentions the importance of respecting diverse cultures, but its primary emphasis is on a global approach. If climate change is a problem threatening the entire planet Earth, then all human beings should work together in response.[26]

The same approach is taken in most serious political efforts on climate change. For example, the United Nations Framework Convention on Climate Change gathers leaders from every nation each year in hopes of negotiating further and more binding agreements. This organization's founding document begins with an assertion that climate change is a global problem, acknowledging "that change in the Earth's climate and its adverse effects are a common concern of humankind."[27] The international framework is offered as a product of and path toward a more global community.

A practical argument for such global attention comes from the communications expert and activist George Marshall, who draws on evolutionary biology to make the case for global moral attention. He argues that human beings evolved to deal with threats to their immediate kin. This poses a challenge: As long as people in Chicago do not feel any sense of immediate community with those in Kivalina, Alaska, they run the risk of viewing the worst effects of climate change as distant and therefore as not urgent.

Marshall recounts a conversation he had five months after Hurricane Sandy with the mayor of Sea Bright, New Jersey. The town was still only beginning to recover from the storm, and two-thirds of its residents were still homeless. Noting that climate change causes more severe and damaging hurricanes, Marshall suggested that the mayor should partner with other towns and cities threatened by climate change and demand federal action. The mayor rolled her eyes and said that climate change "is bigger than anything we could make a difference on. We just want to go home, and we will deal with the lofty stuff some other day."[28] Marshall concedes that those facing immediate threats will focus on their local needs, but he insists that other people need to learn to take on "the lofty stuff" threatening their neighbors across the country—people need to learn to think globally.

Marshall insists that people only take action to defend a community when they feel like a part of it: "If we feel an affinity with the group, then we will willingly make a contribution to prove our loyalty. In times of conflict, we may even sacrifice our life."[29] The more people understand that all other people also share one planet, that they are all part of one human family that deserves everyone's loyalty, the more motivated they will be to act.

### How Big Is the Globe?

George Marshall's moral attention is broad, but not the broadest. He limits moral concern to human beings, resisting environmentalists' efforts to also advocate explicitly for other species. His argument is pragmatic—noting, for example, that the polar bear is "an animal that could not be more distant from people's real life." So, he says, moral appeals that focus on the plight of the polar bear fail to convince anyone new to care or to act.[30] For Marshall, responding to climate change is about creating a global sense of human belonging, and this means leaving out nonhuman creatures.

Of course, Marshall is not arguing that polar bears do not matter. However, he believes that the best way to get others to care about climate change is to talk about human beings rather than any other species. People can rationally understand that all species are connected, that other creatures matter. But, he argues, such claims make no appeal to the emotions that actually motivate action and behavioral change. He is confident that people can learn to care about the universal human family, but he is more skeptical that they can scale up further to a concern for all creatures. Climate ethics requires a balance between the fact that all creatures are interconnected and the countervailing fact that human beings have evolved with a limited moral concern focused on those like themselves.

Naomi Klein suggests a similar argument in her documentary film *This Changes Everything*. She begins the film with a confession, telling the audience, "I've always kind of hated films about climate change. What is it about those vanishing glaciers and desperate polar bears that makes me want to click away?" She returns to this theme toward the end of the film, explaining that polar bears "still don't do it for me. I wish them well. But if there's one thing I've learned, it's that stopping climate change isn't really about them, it's about us."[31] So Klein's film focuses on the human costs of climate change and the movements emerging from human communities to protect their lands, their rights, and their health. This is a common decision in many contemporary discussions of climate change.

However, in the long run, the best way to extend people's moral attention outward, to get them to feel solidarity with all other human beings, might be to extend beyond the human species, to learn to feel solidarity with all beings. Pope Francis, for example, insists that "misguided" and "tyrannical" anthropocentrism contributes to the problem of climate change, and he emphasizes that all creatures are made and loved by God and therefore deserve human care.[32]

The philosopher Kate Rawles rejects anthropocentrism even more resoundingly. She calls for a "Copernican revolution in ethics" that will recognize human beings are not the center of moral concern, just as Copernicus recognized that the Earth is not the center of the known universe. Rawles insists that limiting people's moral attention to humanity ignores the true violence of climate change, which includes "the suffering and death of individual sentient animals in their billions; the widespread extinction of other species; the degradation of extraordinarily complex ecosystems." Once people recognize the wide and precious natural world beyond human experience and knowledge, she argues, they will see climate change not only as a human problem but also as a problem for "a vastly complex and interrelated ecological community."[33] Climate justice, she argues, requires an ethics bigger than any one species.

Pope Francis and Kate Rawles call for a moral community that extends beyond humanity. George Marshall and Naomi Klein note that such an extension of ethics comes with a practical cost, and so they emphasize the importance of human beings over other species. They all agree on the need for a more global morality, a broader sense of community, but they answer the complex questions of scale in different ways.

## Crossing Scales: Toward a Planetary Climate Ethic

It is also important to consider the possibility that morality does not need to scale upward, does not need to be broader. Perhaps concerned people do not need to learn to think "bigger" but instead need to adapt to the fact that every person comes from a particular and specific place that inevitably defines her or his thinking and being in the world.

The theologian Whitney Bauman offers a caution against "global" responses to climate change, arguing that it was globalization that created the problem in the first place. Europeans carelessly expanding their impact on other continents inaugurated centuries of expansionist colonization. Inventors and businesspeople carelessly expanding their impact on the natural world created the industrial system that makes environmental degradation and overconsumption possible. In these historical examples, the impulse was positive—to explore the world, to share true faith, to increase wealth—but the results were destructive because of myriad unintended consequences. A "global" ethic, Bauman worries, indulges an age-old temptation to assume that one can understand the "other," that other people or other species can be viewed as versions of one's self.

Bauman is also suspicious that the kind of unity imagined by many global thinkers could ever really be possible. He does not believe that all humanity or all creatures could ever be a stable, meaningful community. Drawing on queer theory, he points out that "concepts are permeable and identities are always being constructed." Just as there should be no single, normative way to express human sexuality, there should be no single, normative way to be human, or to be a creature. As people and creatures develop and evolve, their roles and memberships in communities will change, and so the communities themselves will inevitably change. This requires flexibility and multiplicity rather than monolithic global truths. Bauman does not hope for a united global community but instead a myriad of options, "an evolutionary rainbow of possible planetary becomings."[34]

The alternative to global morality, for Bauman, is "planetary" morality, which embraces the wide diversity of life on planet Earth in each of its diverse local expressions.[35] He argues that the challenge of climate change should not be met by reminders about commonality—"one human family"—nor by an attempt to expand human communities—"an integrated ecological community." Instead, he suggests, climate change calls for a morality that celebrates differences, caring for others not because one belongs to the same group but rather precisely because one does not. Seeing the world as one human family or one community of creatures simply extends identity politics, indulging people's impulse to only take care of those like them. Bauman hopes that concerned people can, instead, learn to care for and defend others in all their rich uniqueness.[36]

Although global morality seeks to expand outward, planetary morality acknowledges that each person is located in and limited to particular places. It seeks not to be more encompassing but instead to be more accepting. Planetary morality embraces diversity and responds to others' needs simply because they are other: "We are male, female, heterosexual, homosexual, bi, trans, queer, black, white, brown Latino/a, American, Japanese, Kenyan, and more generally of specific descents, yet we are also animal, biological, planetary, and ecological, inextricably bound within and to the planet Earth."[37] All share a fate on this planet, but in very different ways. From a planetary perspective, the differences need to be emphasized and embraced far more than the commonalities.

A planetary ethic is multiscalar. It calls climate activists to be fully present to the immediate world as it is, in all its diversity—at the global, national, and local levels. Planetary thinking can acknowledge that the problem of climate change is global, but it suggests that there will never be one global answer.

Planetarity would mean caring for the residents of Kivalina not because they are human just like me but instead because they are a unique expression of humanity, melding the traditions of the Iñupiat with contemporary US culture in all its complexity. This changes how concerned people might respond to their plight; whereas a global ethic would ask how to pay for the Kivalinans to transition to another place, a planetary ethic insists upon the deeper task of empowering them to not only move but also continue their unique cultural expressions.

A planetary ethic also suggests that people should learn to care about polar bears precisely because polar bears are so different from humans, because they represent an entirely different experience of the world. Difference is to be celebrated, and so the differences between creatures matter. The previous section discussed a tension between George Marshall, who argues that it is too much to expect moral thinking beyond the human species, and Kate Rawles, who argues that the root of the environmental problem is people's selfish attention to solely human interests. The disagreement between them is how big moral thinking should be—can people expand beyond one species, can they morally appreciate their common ground with all other creatures? Planetary morality shifts away from this question by changing the discussion away from the size of moral attention. In this perspective, there is not merely one moral sphere that either includes or excludes others. Instead, morality is about how people deal with others, with that which is not like themselves. So, a planetary ethic asks: How can people learn to care for that which is different from themselves, both within and outside the human species? Humanity is not one community but many. Other species are not one category but many. The task is to learn to embrace difference in all its forms.

And, yet, concerned people cannot do everything at once. We can work to embrace difference, but how will we balance the immediate needs of polar bears and the residents of Kivalina? Where should our moral attention be devoted first? This is one of the wicked challenges of climate change—a place where we should look for help from the witness of Jane Addams.

## A MULTISCALAR RESPONSE
## TO CLIMATE CHANGE

Ultimately, I see Jane Addams's morality as best aligned with the planetary scale of attention because she celebrates diversity and local thinking as well as broad community. Drawing from her work, in this section I argue that global morality is too unidirectional—if people only seek to expand their attention, then they risk missing the complexities of particular places. Drawing on her pragmatism, I also argue that though moral attention to nonhuman creatures is a vital ultimate goal, it is not immediately important to insist upon it. The power of planetary morality is that, like the witness of Jane Addams, it is multiscalar and adaptable.

## The Local in the Global

Jane Addams was very aware of global problems, and so she worked to prevent wars, to form international organizations, and to empower the poor in every corner of the world. She learned to think beyond her own community, and she emphasized the common ground between wealthy people like herself, the poor immigrants in her neighborhood, the foreign soldiers waging war against her nation, and every mother, father, and child around the world. Although she did not use the word "global," she frequently aspired to "universal" thinking. For example, she argued that the privileges of wealthy people "must be made universal if they are to be permanent" and that the task of religion is "to lift a man from personal pity into a sense of universal compassion."[38]

Despite these universal aspirations, Addams's work as a whole offers a corrective to any approach to climate change that is *solely* global. Because it attempts to be so encompassing, global thinking is necessarily abstract— to say that every human being is a member of one family is to conjure a fairly intangible idea of what it means to be "a family." Addams always began from concrete rather than abstract ideas. She advocated democracy as an ideal, but she always emphasized the basis of democracy in local politics and elections. She asked her audience to think about a fight between two boys before scaling upward to consider continental warfare. She never fully trusted her ability to represent others with different experiences, and so she invited a neighbor to be with her when she would speak about poverty in Chicago. She therefore offers concerned people resources to ground global morality in local attention.

A moral response to climate change must include global attention— must acknowledge, with Pope Francis, that the problem involves "every person living on this planet." Meaningfully responding to the challenges facing our human neighbors in Kivalina, Sudan, and Bangladesh requires expansive attention. However, the authority Addams drew from specific experiences in her neighborhood, from knowing one specific community in all its local complexity, suggests that such global attention will not be enough by itself.

When Addams wrote about the children of her neighborhood pretending to "kill Spaniards" during the Spanish-American War, she was making an argument about the importance of global events for local communities. She suggested that no one who wishes to make peace on the streets of Chicago can ignore a war on the other side of the world because the two are connected. But she then quickly moved to discuss the effects of a national strike on her neighborhood, observing that owners lined up against workers could also turn into "prolonged warfare," with an equally dangerous influence on children's mentalities. The rhetoric of a strike could too easily nurture "mutual hate" and become "a menace to social relations."[39] She viewed international violence, urban violence, and neighborhood violence all on a single continuum.

In part, this means that every person of privilege should learn about the effects of climate change on our own neighborhoods, including the local threats of natural disasters and/or scarcity. We should reach out to those around us who may already be suffering from extreme weather, reduced water supplies, or natural disasters.[40] To meaningfully care about how climate change threatens the whole "human family," concerned people must learn to care about specific families and specific persons. Jane Addams demonstrates that global morality needs the richness and complexity of local attention. This is an important lesson for anyone concerned about climate change.

## Pragmatically Valuing Other Species

Jane Addams's witness offers tools to support moral attention beyond human beings, but I think it also suggests pragmatic doubts about the priority of such work. One of the key events that motivated Addams to found Hull House was the experience of cruelty at a bullfight, and so it is clear that she believed any form of violence, including violence against nonhuman creatures, warps the human spirit. At one point, she defined the settlement movement according to its commitment "to insist upon the unity of life," and she understood better than most people in her time that human well-being depends upon healthy environments and relationships with other species.[41] Thus, it is sensible to use her witness to justify an even more expansive moral argument, accounting for the ways climate change is endangering the lives of polar bears in the Arctic and encouraging the growth of invasive species that threaten others around the world. One could, therefore, imagine Addams resonating with the call of the philosopher Kate Rawles for a "Copernican revolution" removing human beings from the center of moral consideration.

However, when Addams argued that every voice should be valued and heard, she tended to limit her moral argument to human beings—to those who could enter into political, economic, social, and cultural conversations. Learning from Addams can raise questions about the pragmatic value of too broad a scale of moral attention in a world where people are not yet good at relating to their neighbors, much less other cultures or other species.

In 1912 Addams helped to formulate a platform for the Progressive Party, which was founded by Theodore Roosevelt as he sought a second presidential term. She helped to craft a set of policies that would have given women the right to vote, limited the influence of lobbyists on elections, and extended the social safety net—all goals in which she deeply believed. However, the platform also called for the United States to build two new battleships a year, conflicting with her pacifist principles. She reflected, "I confess that I found it very difficult to swallow those two battleships." But she nevertheless decided to support the platform as a whole, believing that the Progressives were "most surely on the road toward world peace," despite their imperfections.[42]

This reflects Addams's pragmatic style; she knew her principles and stuck closely to them, but she was willing to compromise, to prioritize action over perfection. Democracy, as she understood it, is not about always finding the right answer but rather finding the best answer on which people can agree. She never compromised her pacifism, but she was strategic about when she insisted on it. She wrote: "The most unambitious reform, recognizing the necessity for this consent, makes for slow but sane and strenuous progress, while the most ambitious of social plans and experiments, ignoring this, is prone to failure."[43] Small steps in the right direction that have been democratically agreed upon are far more powerful than boldly radical statements that are widely dismissed.

In my view, climate change is caused by anthropocentric habits of thought and behavior, and it is impossible to fully and resolutely resist the violence of environmental degradation without moving toward a broader understanding of the moral community. Unlike Naomi Klein, I am moved by the plight of polar bears and am disturbed by shrinking glaciers. I am also concerned about declining rainforests and expanding kudzu and the plight of industrially raised cows and chickens and countless other effects on the nonhuman world. To me, climate change remains both a disastrous environmental problem and a human problem. However, I learn from Addams that it is more important to take positive steps forward than to insist on moral absolutes. The Copernican shift in ethics is an important goal, but no one should expect a moral revolution before beginning the work of resisting climate change.

Many people struggling for climate justice join George Marshall and Naomi Klein in being relatively unmoved by the plight of nonhuman creatures. To change their minds and build a movement around the plight of polar bears, native species, and cows will be important work for the future. However, human rights activists, native rights activists, climate hawks, and many others can all agree now that what is happening to coastal communities around the world is deeply problematic and demands a response. I do not need to convince them all that chickens and polar bears deserve moral attention to join them in resisting the violence against the people of Kivalina. Climate change is a big and wicked enough problem that there is no one approach to solving it, and no action should be held up by a standard of moral perfection.[44]

Ultimately, a nonviolent response to climate change calls for moral attention that includes all creatures. But, like Addams's pacifism, those of us who believe in such a goal must recognize that we can work toward it while also collaborating and cooperating with others who do not share it. The moral value of the nonhuman world is important, but it need not be the most important or primary question right now.[45]

## Toward a Planetary Morality

Jane Addams has much to teach those who advocate a global morality in response to climate change, whether advocating for a human family or a community of all creatures. However, my reading of her work suggests that she fits best with the planetary scale of attention, crossing multiple scales to argue against any single global family, embracing instead the diversity and complexity of evolving identities and communities.

As the theologian Whitney Bauman makes clear, a key advantage of planetary rather than global morality is adaptability. Although global approaches emphasize the need for a common moral vision and must therefore seek universal agreement on a single project, planetary approaches are open to multiple possibilities at once. Planetarity allows different peoples to try different strategies, accepting that there is no single right answer.

Jane Addams regularly praised adaptation and diversity. In *Twenty Years at Hull House*, she wrote: "The one thing to be dreaded in the Settlement is that it lose its flexibility, its power of quick adaptation, its readiness to change its methods as its environment may demand."[46] She had helped to found the house as an educational project where the privileged would serve the needy, but she quickly adapted as she learned from her neighbors. Hull House became the site of a wide range of programs in which diverse members of the community served as equals, teaching and empowering one another though their differences. She never sought to make her neighbors more like herself but rather to help them to be themselves while she remained herself.

Addams also demonstrated this commitment to diversity and adaptability as she scaled her attention upward to national policies and international agreements. She changed strategies and rhetoric as the nation became less accepting of her pacifism during World War I and more accepting of it afterward.[47] She rejected vague appeals to a "love of humanity" throughout her life, and she was more interested in creating concrete relationships in all their complexity.[48] She engaged and celebrated differences rather than seeking to abstract common ground. Although she was aware of the importance of global ideals, the ways she pragmatically brought them into conversation with local specificities suggest a more planetary approach.

Privileged persons concerned about climate change should think in dialogue with Jane Addams about how to work toward planetary democracy, how to value all voices and all perspectives, how to empower as many people as possible to have a voice and a role in shaping their future while respecting differences and engaging in particularities.

## A PLANETARY PROBLEM

The witness of Jane Addams offers insight into a planetary response to climate change, one that crosses scales and values the diversity of Earth's communities. But there are, still, no easy answers to the violence of climate change.

This difficulty is evident in any discussion of the disappearing Alaskan village of Kivalina. The injustice of what is happening to the people of Kivalina is clear and undeniable. They are already survivors of the white supremacy that displaced native peoples and devalued their culture, and they are now being driven from homes their grandparents were forced to make. Their requests for help from those who bear the most guilt for their plight—fossil fuel companies and the people of the United States—have thus far been ignored.

In the terms introduced in chapter 1, this is violence. Using the ethical principles laid out in chapter 2, it is clear that such violence calls for resistance. Privileged people share in the guilt of climate change and therefore in the guilt of slowly destroying the village of Kivalina. We are called to act, to resist violence. But how? What would a nonviolent response to Kivalina be?

Although there is no easy answer, the planetary morality derived from Jane Addams in this chapter offers a few guidelines. First, Addams saw every moral challenge as, in part, an educational challenge. Before the United States can address any problem, its citizens must learn to be concerned. Those of us who have some sense of what is happening in Kivalina should tell others and should raise the issue in political debates. This will not solve the problem in the immediate term, but it will at least help people to face the violence of climate change, perhaps contributing to a more honest national debate.[49]

Second, Kivalina calls all concerned people to local action. Just as Addams learned to be an international peacemaker by engaging her Chicago neighborhood, we will be best equipped to understand and empower the residents of Kivalina if we relate this problem to concrete and immediate challenges in our own communities. Those who live far away from Kivalina may not have practical or productive ways to form relationships with its residents, but perhaps we can better understand what is happening to the Kivalinans if we work to support the indigenous peoples in our own regions, whether or not their survival is directly threatened by climate change.[50] For example, I can better argue that Kivalina should be treated justly if I also advocate in concrete ways for justice, treaty rights, and respect for the Coast Salish peoples of the Pacific Northwest.

A related way to morally engage the challenge of Kivalina is to better understand the lives of anyone who has been made homeless by forces beyond their control in one's own neighborhood. Those of us who are concerned should engage these homeless people, help to empower them, interrogate the systems that ignore them, and support organizations that advocate on their behalf.[51] Such local efforts will not directly save the residents of Kivalina, but they will help us to better understand and care for real human beings in all their complexity. Because violence is an interconnected network, it is productive to do what one can where one can.

If concerned people learn about the importance of democracy from Jane Addams, then no response to Kivalina will be complete without insisting that its residents must have the right to speak up for themselves and to determine their own future. They should be heard in the courts and in Congress. Privileged people cannot and should not speak for the people of Kivalina—they have suffered too long from other people deciding what is best for them. However, we can speak up for their right to be heard, for a political process that takes them and their interests seriously. They should have the right to speak because they are human beings. They should have the right to speak because they are US citizens. Perhaps most important, they should have the right to speak because they are members of a unique culture that has made a life on an arctic island despite a legacy of oppression and violence. Those of us who are privileged have the power to demand that our elected representatives take the Kivalinans seriously. We should keep insisting until someone listens.

This conclusion is unsatisfying. Given the realities of my nation and my life, I have no proposal that will save Kivalina. But I am committed to continuing my attention to the problem, to educating others about it, to asking those in power how they will respond, and to helping those without homes or rights in my own neighborhood. This is the best witness I can give in the face of grave injustice.

Jane Addams responded to violence by devoting her life to democracy. She insisted that every person matters, because she believed that no one can ultimately thrive unless everyone does: "Our feet are mired in the same soil, and our lungs breathe the same air." To resist the global violence of climate change, one must seriously study the soil in which one stands and get to work with one's neighbors cleaning the air.

# NOTES

For biographies of Jane Addams, see especially Elshtain, *Dream of American Democracy*; Brown, *Education of Jane Addams*; and Knight, *Jane Addams*. For analyses of Addams's thought and importance, see especially Fischer, Nackenoff, and Chmielewski, *Jane Addams and Democracy*; Hamington, *Social Philosophy of Jane Addams*; and Schneiderhan, *Size of Others' Burdens*.

1. On the practical work and politics of the "garbage wars," see especially Knight, "Garbage and Democracy."
2. Addams, *Democracy and Social Ethics*, 120.
3. Addams, *Twenty Years at Hull-House*, 77–78.
4. Knight, *Citizen*, 190. How deeply Addams held her faith is a matter of some scholarly debate. It seems safe to say, however, that Addams's religious beliefs and practices, like everything else about her, continued to develop over the course of her life. Though she rarely discussed her own faith and had relatively low attachment to the Presbyterian Church or any particular congregation, she never publicly distanced herself from Christianity, and she used Christian ideals and concepts throughout her public career. See especially Stebner, "Theology of Jane Addams"; and Brown, "Sermon of the Deed," 21–39.
5. Addams, *Jane Addams Reader*, 25–26; also see 17.
6. Addams, *Second Twenty Years at Hull-House*, 404.
7. Addams, *Twenty Years At Hull House*, 175.
8. Ibid., 82. On this aspect of Addams's education, see Brown, *Education of Jane Addams*, chap. 13.
9. See especially Elshtain, *Dream of American Democracy*. Addams did help to raise a niece and nephew after their mother died, and she had many important relationships with many other children. Though she never married, she had a lifelong romantic friendship with Mary Rozet Smith, a Hull House supporter. See Knight, "Love on Halsted Street," 181–200.
10. Addams, *Newer Ideals*, 5.
11. Addams, *Writings on Peace*, 285. For more analysis of Addams's ideas about peace as an evolutionary possibility, see especially Green, "Lessons from Jane Addams," 223–54.
12. Addams, *Democracy and Social Ethics*, xi.
13. Addams, *Twenty Years at Hull House*, 75.

14. Ibid., 206. See also Charlene Haddock Seigfried, "Introduction," in *Democracy and Social Ethics*, by Addams; and Fischer, *On Addams*, chap. 2.

15. Addams, *Jane Addams Reader*, 62.

16. Addams, *Democracy and Social Ethics*, 7, 117.

17. Ibid., 93. Interestingly, she made the same argument about child labor in chapter 5 of *The Spirit of Youth and the City Streets*. This chapter exemplifies Addams's pragmatism; she almost certainly would have preferred that fourteen-year-old children be in school rather than in factories, but she knew that the practical realities of many poor families required that even young children work. She acknowledged this reality and insisted that, for example, a child in a sewing factory be taught that "the design she is elaborating in its historic relation to art and decoration if she understands "her daily life is lifted from the drudgery to one of self-conscious activity, and her pleasure and intelligence is registered in her product." Addams, *Spirit of Youth*, 122.

18. Addams, *Newer Ideals*, 79. Addams also made the connection between her neighborhood and international conflict in *The Second Twenty Years at Hull House*, which argued that her settlement's story could not be told apart from the story of World War II: "It is idle to speculate on what an infinitesimal unit like Hull-House or any other of the millions of units composing the social order, would have been like if the world war had never taken place. But whether we are for it or not, our own experiences are more and more influenced by the experiences of widely scattered people; the modern world is developing an almost mystic consciousness of the continuity and interdependence of mankind. There is a lively sense of the unexpected and yet inevitable action and reaction between ourselves and all the others who happen to be living upon the planet at the same moment." Addams, *Second Twenty Years at Hull House*, 7.

19. Addams, *Twenty Years at Hull House*, 153.

20. Addams, *Jane Addams Reader*, 337. Maurice Hamington articulates the approach well: "Addams's response to war in Europe parallels her response to problems that arose in the Hull House neighborhood: mobilize sympathetic knowledge and a community of activists to search for rational and caring solutions leading to lateral progress" Hamington, *Social Philosophy of Jane Addams*, 90. See also Green, "Lessons from Jane Addams."

21. As Erik Schneiderhan explains: "Addams created boundaries. She was frequently ill and took time to convalesce without working, often for months at a time. Further, she did not use *all* of her money on Hull-House; she maintained a standard of living that included well-appointed housing, frequent travel, and exposure to culture through books, fine art, music, and theater." Schneiderhan, *Size of Others' Burdens*, 49.

22. Addams, *Twenty Years at Hull House*, 152.

23. See especially "Kivalina: Village Profile," http://nana.com/files/pdf-bios/NANA
-VillageProfile-Kivalina.pdf; and Shearer, *Kivalina*.

24. For my previous analysis of scale as an issue of Christian ecological ethics, see
O'Brien, *Ethics of Biodiversity*, chaps. 4 and 5.

25. This idea that we need to expand our morality is a common, although not universal, claim in environmental and ecological ethics. See especially "The Land
Ethic" in *Sand County Almanac*, by Leopold.

26. Pope Francis, *Laudato Si'*, §3, 164.

27. United Nations, "Framework Convention on Climate Change," 1992, http://
unfccc.int/files/essential_background/background_publications_htmlpdf
/application/pdf/conveng.pdf.

28. Marshall, *Don't Even Think About it*, 7.

29. Ibid., 196. He also writes: "Climate change is the one issue that could bring us
together and enable us to overcome our historic divisions" (pp. 229–30).

30. Ibid., 128, 137.

31. Lewis and Klein, *This Changes Everything*.

32. Pope Francis, *Laudato Si'*, §68, 118.

33. Rawles, "Copernican Revolution in Ethics," 90–91.

34. Bauman, *Religion and Ecology*, 8.

35. He defines the terms this way: "Whereas globalization is the imposition of same-
ness over the face of the planet, planetarity is a way to think about how we are
codefined and come together because of, with, and through our differences."
Ibid., 209.

36. Willis Jenkins makes a similar argument in *The Future of Ethics*. Although he
uses the language of "global ethics," I interpret his approach to be compatible
with what Bauman labels "planetarity." Consider chapter 3, which argues "for
developing global ethics from the cross-border practices of agents collaborating
to respond to planetary problems. Its more important products are not documents and declarations, but the moral creoles and middle axioms generated
from pluralist reflection on shared problems" (p. 141).

37. Bauman, *Religion and Ecology*, 73.

38. Addams, *Jane Addams Reader*, 17.

39. Addams, *Newer Ideals*, 80. Of course, Addams did not make this argument to
suggest that strikes should never occur or that striking workers were always
wrong. Instead, she urged all sides to prevent strikes when possible and to
ensure, when a strike is necessary, that civility and peace be given priority.

40. For attempts to chronicle and communicate the ways daily weather experiences
in particular places reflect broader climatic shifts, see www.wunderground.com
/climate/local.asp and http://thealmanac.org/. I do not know of a similar project
that focuses on issues of climate justice.

41. Addams, *Jane Addams Reader*, 59.

42. Addams, *Second Twenty Years at Hull House*, 35.

43. Addams, *Democracy and Social Ethics*, 68.

44. As discussed in the introduction, I would not label my overall approach to climate justice exclusively "pragmatic," but this argument is nevertheless consonant with the move toward "environmental pragmatism" and "moral pluralism" in philosophical environmental ethics. See especially Stone, *Earth and Other Ethics*; Norton, *Toward Unity among Environmentalists*; and Minteer and Manning, "Pragmatism in Environmental Ethics."

45. Using language that Richard Bohannon and I have explored in another text, I am arguing that, in the near term, we should prioritize the specific, human complaints of environmental justice over the idealized universal community eco-justice. See "Saving the World (and the People in it, Too): Religion in Eco-Justice and Environmental Justice," in *Inherited Land*, ed. Bauman, Bohannon, and O'Brien, 171–87. For an excellent account of Christian ecological ethics that explicitly prioritizes eco-justice, see Rasmussen, *Earth-Honoring Faith*.

46. Addams, *Twenty Years at Hull House*, 95.

47. For an analysis of how Addams's views of peace developed over her life, see Green, "Lessons from Jane Addams," 223–54. For a biography that emphasizes Addams's constantly adaptive, developmental thinking, see Brown, *Education of Jane Addams*.

48. Wendy Sarvasy argues that Addams was uninterested in abstract love but passionate about "forging new interdependent relationships . . . between individuals in all their complexity and specificity." Sarvasy, "A Global 'Common Table': Jane Addams's Theory of Democratic Cosmopolitanism and World Social Citizenship," in *Jane Addams and Democracy*, ed. Fischer, Nackenoff, and Chmielewski, 189.

49. For one effort to educate the public about and chronicle Kivalina's relocation, see www.relocate-ak.org/projects/.

50. See especially Harvey, "Dangerous 'Goods.'"

51. See, e.g., Stivers, *Disrupting Homelessness*. This is not a book directly about climate change, but it offers tools for genuine engagement with homeless neighbors, with the systems that cause homelessness, and with efforts to change those systems. These are important lessons for anyone concerned about climate change.

# 5

## Dorothy Day and the Faith to Love

> To work to increase our love for God and for our fellow man (and the two must go hand in hand), this is a lifetime job. We are never going to be finished. Love and ever more love is the only solution to every problem that comes up.
>
> —Dorothy Day, *By Little and By Little*

During the Great Depression, workers across the United States went on strike to demand fairer pay and shorter workweeks. Owners changed laws to make picketing illegal, and so strikers and their supporters were frequently taken to jail. One Saturday in 1935, Dorothy Day and a group of students from a local Catholic school protested in solidarity with striking workers at Ohrbach's department store in New York City. Day and her companions waved signs that quoted Pope Pius XI's emphatic statements in support of organized labor and workers' rights. In her autobiography, Day describes the impact of this protest: "The police around Union Square were taken aback and did not know what to do. It was as though they were arresting the Holy Father himself, one of them said, were they to load our pickets and their signs into their patrol wagons."[1]

This story is typical of Day's public life and activism; she took a strong stand in a political cause but used her deep religious faith to disrupt the expectations of those on both sides of the protest. In the 1930s, most people assumed that strikers were atheist radicals, and most police assumed that they were doing their religious as well as their civic duty by arresting such disruptive rabble. But a proudly Catholic protester disarmed these assumptions.

Reflecting on how she sustained her political and religious commitments in tough times, Day credited her love for God and other people, insisting that "love and ever more love is the only solution" to every problem. For her, love was the common ground between radical politics and religious faith. She devoted her life to serving the poor, and her faith sustained her in the work that she knew was "never going to be finished"—the work of resisting violence in all its forms.[2]

The wicked problem of climate change raises deep questions about politics, economics, culture, technology, and lifestyle. Day offers a reminder that these questions also all have religious dimensions. Thus, religious people should resist climate change, and the movement for climate justice should identify the sources of faith and love that can sustain it. Using Day as a witness sheds light on the relationship between faith and activism and on how to help today's diverse communities concerned about climate change reflect on what role religious traditions can play in a multicultural twenty-first-century movement.

Day also reveals a very personal reason to think about religious faith. Resisting climate change is a difficult, lifetime job, so contemporary activists need to ask what can empower them to continue the struggle. Some might be inspired to share the Christian ideal of love that strengthened Dorothy Day; others might seek a different motivation. Her witness suggests that all of us who are concerned about climate change need a faith in something.

## HOSPITALITY AND PROTEST

Dorothy Day was a writer. She was the daughter of a journalist, and she wrote for newspapers her entire adult life. She saw journalism as a form of activism, and she hoped to inspire her readers to change their lives and their societies once they understood the realities of poverty and injustice that she wrote about.

For the first ten years of her professional life, Day wrote articles for a series of socialist publications, advocating and supporting a new economic system during this time before the Cold War, when socialism was a common commitment for progressive political activists. At the age of eighteen years, Day moved to New York City and applied to the daily socialist newspaper *The Call*. The editor was impressed by her application but could not afford to pay her a full salary. She offered to start for just $5 a week, the wage earned by many factory workers. The editor agreed, and some of her first published work was about the experience of living at a poverty wage.[3] Never concerned with material wealth, Day felt driven to understand and to help readers to understand the plight of workers.

Two years later, in 1917, Day extended her activism beyond writing into civil disobedience. She joined a suffragist protest in Washington, picketing the White House to insist that women deserved the right to vote. She and her compatriots were trespassing illegally, and so they were arrested and sentenced to thirty days in prison. Twenty years later, she wrote about the dehumanization of imprisonment, but she also realized that she had the comfort of knowing that she would be there only a month and was supported by a movement outside the prison. She felt deep sympathy for the women she met who faced far longer sentences with far less external support and who in many cases had been imprisoned for crimes committed to meet the basic needs of their families. In prison, she learned to care less about her own rights and more about "those thousands of prisoners throughout the country, victims of a materialistic system."[4]

In her prison cell, Day found comfort in reading the Bible. This continued an important but inconsistent thread of her young life, which was characterized by bouts of passionate faith, including a baptism that confused her nonreligious parents and siblings. However, as a teenager she had become steadily more interested in radical politics and less interested in religion. The socialist circles in which she traveled tended to be atheistic, dismissing religion as a comfort necessary only for the weak or a tool of oppression supporting unjust structures.

After leaving prison, Day continued to write for socialist papers and sustained her activism while she trained as a nurse, was briefly married, lived in Europe, and wrote a novel. With the proceeds from that book, she bought a small house on Staten Island in 1925. There she lived with Forster Batterham, a naturalist whom she called her "common-law husband" and with whom she had a baby, Tamar Teresa, in 1926. Day's daughter led her back to organized religion—soon after giving birth, Day approached a nun about baptizing her daughter; and by the end of 1927, she herself had also become a Catholic.

Batterham was an atheist who believed that rational science should replace religion. He could never accept Day's faith, and they separated soon after her conversion. For Day, this was somewhat ironic, because it was life with him and their daughter that had led her to believe in God. She writes: "It was human love that helped me to understand divine love. Human love at its best, unselfish, glowing, illuminating our days, gives us a glimpse of the love of God for man." Having felt the affection of a partner in life opened Day to greater love. Having given birth, she was "awed by the stupendous fact of creation" and filled with love. She became convinced that love is the most powerful force in the universe, and she found resonance for this idea in the Catholic tradition, to which she then devoted herself.[5]

However, it was also important for Day to insist that her conversion to Catholicism did not mean a conversion away from radical social action or advocacy for the poor. Indeed, her first autobiography, *From Union Square to Rome*, is addressed to her brother, who shared her socialist politics but could not understand her religion. She insisted that her faith was an even fuller expression of concern for the poor than her past socialism. Although socialists aspire to help the poor by creating an international movement, she argues, her Church was already an international movement made up of poor people, motivated by love: "The Catholic Church is the church of the poor, no matter what you say about the wealth of her priests and bishops. [The people in church] were of all nationalities, of all classes, but most of all they were poor."[6]

In 1933 Day met a fellow Catholic radical, Peter Maurin, who helped to set the course for the rest of her life. Maurin was a French peasant and a self-educated theologian who believed that Christians are called to radical witness and hard work on behalf of the poor.[7] He inspired Day to create a newspaper, *The Catholic Worker*, in order to share their ideas. The paper developed a readership, and it soon became a weekly, with a circulation of more than 100,000 copies. It continues to be published from New York City to this day.

One of Maurin's ideas was that the Church should run "houses of hospitality" for all who needed food and shelter, and he wrote about this in the newspaper. Hungry and homeless readers began to come to the paper's office to ask where they could find such hospitality. For a few months, Day told them she could not help, that she was focused on writing and publishing; but this changed in December 1933. After hearing the story of a homeless woman who, overcome by hopelessness, committed suicide, Day immediately rented a second apartment and opened it to anyone in need.[8] This house of hospitality soon expanded to increasingly larger properties in New York City, helping more and more people.

This combination of journalistic advocacy for the poor and open hospitality to meet their immediate needs became a movement, the Catholic Worker, which took its name from that of Day's newspaper. This movement spread as the paper's readers started new houses of hospitality in other cities. By the 1940s there were thirty Catholic Worker houses in the United States and one in England.[9]

One of the most radical things about Day and Maurin's movement was its financing. The newspaper's first issue was funded by a few donations, was published from Day's kitchen table, and sold for a penny a copy. She later explained that she needed to charge for the paper in order to get a second-class mailing permit, but she "put the least possible price on it to indicate what [I] feel about money." After writing the second issue, she sold her typewriter to pay for its printing.[10] Slowly, donations came in, but she never changed the price and never sought profits. Day, and those who wrote and edited with her, became "workers" motivated by faith and love rather than money: "We choose to spend the salaries we might be making if we were business-like on feeding and sharing our home with the homeless and hungry. . . . We are willing to clothe ourselves in the donations of clothes that come in, we are willing to eat the plainest and most meager of meals and to endure cold rooms and lack of privacy."[11]

Day's life was shaped by the fact that Jesus Christ—whom she believed to be God incarnate and the founder of her Church—lived in poverty. This pivotal figure in the history of the world, the key signal of God's love for the world, was born into a family of laborers who could not find shelter on the night of his birth. For Day, this signaled that Jesus's followers should refuse the world's standards of success: "Let us rejoice in poverty, because Christ was poor. Let us love to live with the poor because they are specially loved by Christ. Even the lowest, most depraved, we must see Christ in them, and love them to folly. When we suffer from dirt, lack of privacy, heat and cold, coarse food, let us rejoice."[12] Day viewed the depravation of involuntary poverty as an inevitable consequence of an economic system based upon greed, and she resisted it by embracing the virtue of voluntary poverty. She dismissed worldly standards of success in order to live with and understand those victimized by the world.

Day's Catholic Worker movement became widely known not only for voluntary poverty, hospitality for all in need, and journalism on behalf of justice but also for nonviolence. In 1938 Day's column in *The Catholic Worker* clarified the movement's stance: "We are opposed to the use of force as a means of settling personal, national, or international disputes." She argued that there is no way to use violence without becoming destructively violent oneself: "As long as men trust to the use of force—only a superior, a more savage and brutal force will overcome the enemy."[13] This was always a controversial stance, and it became far more controversial during World War II, when most everyone in the United States—including most of Day's fellow radicals and fellow Catholics—supported military action. In 1942 she wrote, "We are still pacifists. Our motto is still the Sermon on the Mount, which means that we will try to be peacemakers."[14] This position shrank the movement, as many Catholic Worker houses closed or changed their names so as not to affiliate with pacifism. The paper's circulation dropped from a prewar peak of 190,000 to 50,000.[15]

After the war ended, the Catholic Workers continued their pacifism, and one of the movement's most sustained actions was a refusal to participate in Cold War–era air-raid drills. During such drills, the law required everyone in New York City to stay indoors, taking shelter to practice for a nuclear attack. Day and other workers refused and instead remained in public squares picketing and insisting that the only true defense against a nuclear attack is to prevent it by reducing the violence in the world. After being imprisoned for this civil disobedience, she wrote in 1957 that she could not "consent to the militarization of our country without protest. Since we believe that the air-raid drills are part of a calculated plan to inspire fear of the enemy instead of the love which Jesus Christ told us we should feel toward him, we must protest these drills."[16]

Day continued protesting for her whole life. She was sentenced to prison at least a dozen times, but considered this a worthy sacrifice for her witness against violence, on behalf of the poor.[17] She traveled around the country and spoke widely about the Christian call to actively love the poor. She became an inspiration for hundreds of Catholic Worker houses and thousands of people who devoted months, years, and entire lifetimes to the movement. She also remained steadfastly Catholic her entire life, attending mass virtually every day. She died in 1980, and her supporters quickly began efforts to have her declared an official Catholic saint, a campaign boosted in 2015 when Pope Francis mentioned her as a "great American" in his speech to Congress.[18]

## "LOVE AND EVER MORE LOVE"

In many ways, Day is more like John Woolman than Jane Addams. Although Addams was pragmatically willing to work with political systems for incremental change, Day and Woolman separated themselves from the violence of the world. Echoing Woolman's desire to be "a fool for Christ," Day wrote that Christian activists must "love to the point of folly, and we are indeed fools, as Our Lord Himself was who died for such a one as this."[19]

Loving others "to the point of folly" kept Day writing about the plight of the poor, feeding the hungry, and protesting against war and oppression for her entire life. Against proposals for gradual reforms and compromises, she insisted that the only proper response to the profound structural violence of the world was "love and ever more love." However, she was not naive. She knew that human beings are hard to love and that love requires sacrifices, and she was daily reminded of these facts by the desperation of her poor and suffering neighbors. She nevertheless committed to love them and everyone else. She was fond of quoting a line from *The Brothers Karamazov:* "Love in action is a harsh and dreadful thing compared with love in dreams. . . . Active love is labor and fortitude."[20]

For Day, the labor and fortitude of love was an act of faith. She believed that "all other loves must be a sample of the love of God." Thus, she sought to treat every person as she would treat Jesus because "God sees Christ, His Son, in us and loves us."[21] Religion and the sacrificial devotion of Jesus's love are not polite topics of conversation in most circles, but Day talked about them constantly and told one biographer, "If I have accomplished anything in my life, it is because I wasn't embarrassed to talk about God."[22]

Talking about God and God's love helped Day to avoid the sin of pride, to move outside her own desires and opinions and to empathize with others. Her nascent faith helped her to focus on others rather than herself when she was first sent to jail. Her maturing faith helped her to open her apartment to the homeless and to devote her life to unpaid work. She frequently missed the small pleasures that poverty made inaccessible—she listed "cigarettes, liquor, coffee, candy, sodas, soft drinks"—but noted that she could overcome these "unnecessary desires" when God helped her to see that her small sacrifices brought her closer to those who live in true need and desperation.[23]

Day's faith helped her to change her life, responding to the structural violence of poverty by living among the poor. Just as she was energized by her ability to talk about God, she also insisted upon talking about the poor: "We must talk about poverty, because people insulated by their own comfort lose sight of it." She believed that true social change could only come from those willing to sacrifice for others, and so she lived with and wrote about the hardships and institutional burdens placed upon the poor.[24]

Ever a radical, Day insisted throughout her life that the structural violence of poverty and inequality was so severe that it called for some kind of revolution. However, after her conversion to Catholicism, she distinguished her goal from communist revolution because she believed that no violent revolution could ever possibly heal the violence of injustice and poverty. She sought, instead, "a Christian revolution of our own, without the use of force," which would truly transform the world and help people to better love one another.[25]

In an untransformed world, this commitment to Christian love regularly seemed foolish and impractical. The Catholic Worker house in New York City was constantly short of funds to publish its newspaper and feed the poor, but Day was never practical about money. A frequently recounted story tells of a supporter donating an expensive diamond ring to the movement. Day immediately re-gifted the ring to a homeless woman who had come to the house for a meal. When asked why she had not instead sold it to buy food or basic necessities, she responded, "Do you suppose God made diamonds only for the rich?"[26] She seemed impractical again when the City of New York bought property from the Catholic Worker and, in addition to the price of the land, offered $3,500 in accrued interest on the sale. Day sent the $3,500 back with a letter explaining that Catholic Workers did not believe in collecting interest on money but that, instead, "we are commanded to lend gratuitously, to give freely."[27] A final impracticality was Day's refusal to participate in political as well as economic life; in 1967, she wrote that despite her arrest fifty years earlier for the cause of women's suffrage, she was a committed anarchist who had never voted.[28]

To those who criticized her impracticalities, Day clarified: "We are *not* here to prove that our technique of working with the poor is useful, or to prove that we are able to be effective humanitarians." She dismissed "usefulness" and "effectiveness" as ideals of industrial society, which treated people as statistics rather than beloved children of God. She measured success based on her faithfulness to her ideals rather than any external measure of success: "We feed the hungry, yes; we try to shelter the homeless and give them clothes, if we have them, but there is a strong faith at work; we pray. If an outsider who comes to visit doesn't pay attention to our praying and what that means, then he'll miss the whole point of things."[29]

Day did not feel called to solve the problem of poverty but instead to side with God against it by helping her poor neighbors in the most principled way possible. Although her work fell short of her ideals in many ways—Catholic Worker houses regularly closed, supporters regularly left in anger, and hungry people sometimes went unfed—her beliefs sustained her: "We admit that we may seem to fail, but we recall to our readers the ostensible failure of Christ when he died on the Cross, forsaken by all His followers. Out of this failure a new world sprang up."[30]

Called to love God and her neighbors with every action, Day also believed that she should not judge others, and her profound witness always aimed at critiquing structural violence rather than the people caught within it. She knew that her way of life was not for everyone, and that her "foolish" resistance to the world benefited from the donations and advocacy of others who participated more fully in economic and political systems. She also knew that the Catholic Worker was never fully separated from the limitations of these systems, as she demonstrated when she returned the interest money to the City of New York. She assured city officials that "we are not judging individuals, but are trying to make a judgment on *the system* under which we live and with which we admit that we ourselves compromise daily in many small ways, but which we try and wish to withdraw from as much as possible."[31]

Day's allegiance was not to any political system and not, ultimately, to any institution. Instead, she sought to love God and to love the people God had made. She insisted that God created everyone with love and for love, and so she strived to love everyone. This made her confident that God's love can be expressed outside of any church or faith tradition. In the 1950s, when much of the nation was afraid of communism and avowed communists were being accused of treason, she stood up for them, and she used her religious authority to insist that these atheists deserved respect. She thanked secular radicals who "helped me find God in His poor, in His abandoned ones, as I had not found Him in the Christian churches."[32] Late in her life, she began to question whether it was even useful to distinguish between religious and secular activists: "The longer I live, the more I see God at work in people who don't have the slightest interest in religion and never read the Bible and wouldn't know what to do if they were persuaded to go inside a church."[33]

## RELIGION, FAITH, AND CLIMATE CHANGE

The hard-working love of humanity that Dorothy Day advocated has much to teach the movement seeking climate justice. But it is worth considering whether the religious language Day used to advocate this love—indeed, whether any religious language—is appropriate to the challenge of climate change and worth the baggage it brings along. Should people who want to resist climate change in the twenty-first-century talk about God?

### Climate Change as a Religious
### and an Economic Problem

Many people who are deeply concerned with climate justice are motivated by religious faith, and they argue that this faith is in fact a vital resource for the movement. One argument along these lines is that only religion can stand in opposition to the destructive economic systems that threaten the climate and human communities.[34]

For example, consider the Buddhist philosopher and activist David Loy, who follows in Dorothy Day's footsteps by insisting on religious resistance to unchecked free markets. Loy argues that climate change is caused by a false faith in capitalism. The global market has come to fulfill a religious function for most of the world's people; thus, its goal—wealth—has become the dominant form of salvation embraced by the world community, and its religious practice—consumerism—is believed to provide salvation from suffering and scarcity. Capitalism, as Loy interprets it, teaches that all problems can be solved by "its god, the Market," and fine points of doctrine can be explained by its theology, "the discipline of economics." Indeed, Loy calls free market capitalism "the most successful religion of all time" because of its global influence and unquestioned role in the lives of people in industrial and developed countries.[35]

The problem, for Loy, is that capitalism is a "false" religion, encouraging its followers to treat both people and the planet as commodities to be quantified and consumed rather than beings to be respected or loved. He believes that "the degradation of the earth and the degradation of our own societies must both be seen as results of the same market process of commodification" nurtured by the religion of capitalism.[36] Capitalism fails as a religion, according to Loy, because it is based upon a delusion—that infinite economic growth is possible in a finite world—and a moral mistake—telling people that they should indulge their greed.

The answer to such false religion is truth, and so Loy calls "all genuine religions" to distance themselves from market systems and market beliefs. He seeks a coalition of diverse faiths united to defend the natural world and the poor. In his view, true religions teach generosity and emphasize the interconnectedness of all things, and these ideas will nurture healthier practices and more just and sustainable ways of life. He demonstrates that his own religious tradition, Buddhism, has tools to teach this healthier path and to resist market capitalism, and he is confident that other traditions like Christianity, Islam, Judaism, and Hinduism can do the same.[37]

Loy is less likely to talk about "God" than Day, focusing more on the path and practices of Buddhism, a tradition that is often nontheistict. But this example demonstrates that activists like him are following in Day's footsteps from other religious traditions. Like Day, Loy views capitalism with deep suspicion and seeks to replace it with the teachings of an ancient religious tradition. His call for people to separate themselves from market logic recalls Day's refusal to focus on profits or follow pragmatic economic principles while running the Catholic Worker. Perhaps the best way to resist the changing climate is to resist the market with the resources of traditional religion.

## Toward a New, Environmentalist Religion

An alternative view suggests that traditional religions do not have suffi-cient resources to combat climate change and that in fact those religions themselves must be overthrown if human beings are ever to live har-moniously with the surrounding world. One important contemporary voice in this discussion is that of Bron Taylor, a scholar of religion who expresses concern that the environmentalism of traditional religions is, at best, "indirect," and is unlikely to motivate radical changes for the sake of the planet. Taylor is more interested in newer forms of religion that are emerging among radical environmentalists and outdoor enthusiasts, people who find spiritual connections separate from traditional religion. Their religious impulses come "from a deep sense of belonging to and con-nectedness in nature," in which "nature is sacred, has intrinsic value, and is therefore due reverent care." He calls this more direct spiritual relationship with the natural world "dark green religion."[38]

Taylor's primary scholarly interest in dark green religion is descrip-tive—he has observed a new religious movement with global reach and seeks to understand it. However, in "A Personal Coda" to his book on the subject, he explains the special concern that motivates his studies. He believes the most influential religions in the world are simply too outdated and "light green" to fully respond to contemporary challenges. They are founded in "ancient dreams, . . . for which there is no evidence and many reasons to doubt." So Taylor turns his focus to new religions, which are founded in "the real world" of "an evolutionary-ecological worldview."[39] Presumably, then, he would see limits in Loy's environmentalism insofar as it remains attached to the traditions of Buddhism and a cosmology that includes ancient elements.

Interestingly, Taylor's focus on adaptability and worldliness in envi-ronmental religion leads him to be far less dismissive of market capitalism than Loy. Taylor notes "tantalizing possibilities" of corporations contribut-ing to dark green religion, motivated by changing ideas among their cus-tomers to articulate environmental concerns and environmentally sensitive worldviews. As an example, he cites the Japanese electronics company Sanyo, which has a corporate philosophy that "sees the Earth as a single living organism" and therefore seeks "to create the products needed to help us live in harmony with the planet."[40] Taylor also identifies important dark green impulses in the cultural productions of the Walt Disney Corporation, which produces films like *The Lion King* and *Pocahontas* that nurture "reverence for nature and feelings of kinship with the natural world."[41]

Loy appeals to the traditional religions that he calls "genuine" because he seeks deep traditions that can compete with the increasingly dominant forces of capitalism. By contrast, Taylor is open to the environmentalist impulses of corporations and the market because he is deeply worried about traditions that cannot adapt to contemporary problems and new understandings of reality. For Taylor, genuine resistance to the violence of climate change can and should be fueled by religion, but this religion should be fully devoted to the natural world rather than to outdated ideas and traditions. He might see much to admire in Dorothy Day's commitment to social activism; but in the struggle against climate change, it seems, he would advise less explicit or devout attachment to ancient faiths.

## Rationality Over Religion

Still a third approach separates religion entirely from climate justice. If religions have contributed to mistaken ways of thinking in the past and have a tense relationship with science in the present, then perhaps there is no reason to have religion in the discussion at all.

One version of this argument comes from the oceanographer and former White House science adviser Jeff Schweitzer. His premise is stated straightforwardly in a blog post: "We will not effectively address the issue of global warming if we appeal to religion." Schweitzer particularly emphasizes conflicts between religion and science, and he worries that any attempt to include religious people in climate activism would compromise its scientific basis.[42] Religion, as he understands it, is about unverified faith rather than concrete evidence, and a choice must be made between the two. Climate change, he insists, calls for "the most cogent, fact-based, scientifically sound arguments possible given the evidence in hand. Any deviation from that course is irresponsible."[43]

For Schweitzer, as for many other environmentalists, religion is unnecessary at best and destructive at worst. Like Bron Taylor, he sees traditional religions as too archaic and otherworldly to respond to the contemporary challenge of climate change. Moving even further than Taylor, however, Schweitzer argues that religion in general is backward and, thus, that any spiritual or religious response to environmental problems would be counterproductive. For him, "religious morality has failed." Key evidence for this is climate change itself; existing moral codes are indicted by the facts that people denied the evidence of changing climate and then failed to take meaningful action when they began to understand the problem. Human beings can and should develop a new moral system, "divorced from god and religion," based upon science and reason.[44] The answers to humanity's biggest moral problems will come from those who think rationally rather than spiritually.

While refusing the label of "atheist," Schweitzer emphasizes that he is constructing his life and developing moral principles based on reason: "I am a rationalist, and if others wish to believe in an invisible man in the sky with magical powers, we can label them arationalists."[45] For Schweitzer, any productive and organized action on climate change will need to be based on reason rather than religion, bringing people from different cultures together and overcoming simplistic prejudices to respond constructively to the perils faced by the global community in a time of atmospheric change.

Loy's appeal to "genuine religion," Taylor's appeal to "dark green religion," and Schweitzer's appeal to "rationalism" offer three contrasting positions, and choices must be made between them. Does religion have a role in resisting climate change? And if it does, what kind of role and what kind of religion?

## THE FAITH TO RESIST CLIMATE CHANGE

This book is not the place to objectively consider such questions. Readers already know I assume that religion should be part of the movement for climate justice, and my appeal to explicitly Christian witnesses suggests my appreciation for traditional religions. So it should come as no surprise that here I draw on Dorothy Day's witness to argue that religion is an essential part of resisting the structural violence of climate change.

However, this does not mean a simplistic support for all religion, nor an utter dismissal of new religious developments and rationalist approaches to climate activism. Instead, I seek a movement for climate justice that could welcome all, whatever their approach to religion might be. Day's legacy must be translated for a movement that includes not only religious activists like David Loy but also more skeptical thinkers like Bron Taylor and Jeff Schweitzer.

Day had the zeal of a convert from the day her daughter was baptized until her death fifty-three years later. However, despite her deep personal devotion, she spent little time trying to convince others to believe in God or become Catholic. Instead, she urged people to love one another enough to ensure that their neighbors had enough to eat, warm clothes, and a place to sleep. Her sternest critiques were not directed at nonbelievers but at people who claimed to be Christian but did not feed the hungry or care for the poor. As she wrote, "When we meet people who deny Christ in His poor, we feel, 'Here are atheists indeed.' "[46]

The most important lesson to draw from Day, then, is not that resistance to climate change will require affiliation with any particular religious community but rather that climate activists must have enough faith in something to take the problem of climate change seriously, to oppose the forces causing the problem, and to truly love their suffering neighbors. Some activists may well deny the existence of God, but what truly matters is that they never deny the suffering of others. That would make them what Day called "atheists indeed."

## An Earthly Faith

Bron Taylor is wary of traditional religions because of their tendency to be otherworldly, to focus on a paradise imagined centuries ago—"ancient dreams"—rather than the real world, here and now. Jeff Schweitzer is critical of religion because he sees faith too often used as an excuse to avoid the problem of climate change or deny scientific truths. These are real challenges, and any serious climate activists should agree with Taylor and Schweitzer that religiously informed responses to climate change can be dangerous.

For example, during a 2010 congressional hearing, Illinois representative John Shimkus argued against regulating carbon emissions by appealling to his faith. He said, "I do believe in the Bible as the final word of God, . . . and I do believe that God said the Earth would not be destroyed by a flood."[47] In the book of Genesis, after Noah's ark lands on dry ground, God makes a covenant never again to flood the Earth. Shimkus interprets this to mean that God would never again allow humanity to be harmed by the weather. Trusting in a text written more than two thousand years ago to interpret the twenty-first-century climate system, Shimkus believes that the climate will not change on God's Earth. So he dismisses reports of sea level rise as necessarily inaccurate, and he has worked against any activist response to climate change.

If this were indicative of how *all* people of faith thought about climate change, it would be a profound indictment of religion. But it is not. Many Christians and people of other faiths take scientific evidence very seriously, and thus they are prepared to view scripture and faith traditions in the context of the best available evidence about what is happening to the world. Human activity is changing the atmosphere, and sea levels are rising. These are scientific facts, and religion need not dispute them.

Dorothy Day had very little to say about the relationship between faith and science, but her witness does offer some insight into how Shimkus's faith could be seen as bad theology. His thinking is not grounded in the realities of God's world, not responsive to current events, and not invested in earthly truths. Day believed in an almighty God in heaven, but this belief drove her to pay more attention to the world around her, not less. She profoundly critiqued believers who responded to "spiritual hunger" with prayers and good intentions but ignored material hunger, poverty, and the social forces and structures that made them worse.

In 1940 Day articulated the core vision of the Catholic Worker movement this way: "We are working for 'a new heaven and a new *earth*, wherein justice dwelleth.' We are trying to say with action, 'Thy will be done on *earth* as it is in heaven'" (emphasis in the original).[48] The way to follow the Christian God is "to prostrate oneself upon the earth, that noble earth, that beloved soil which Christ made sacred and significant for us by His Blood with which He watered it."[49] Rationalists and advocates of new religions could certainly question Day's faith that the blood of Christ makes the Earth sacred. But one must admit that, at least in her case, this faith led to further engagement in earthly realities rather than a departure from them.[50]

Day's conversion to Christianity occurred after she moved to the sea-side and lived with a naturalist who taught her about the intricacies of aquatic ecosystems. Later in life, she told a biographer how important it was to occasionally return to the ocean, because it helped her to quiet her mind: "When I look at the sea I know that we are meant to stop our intel-lect dead in its tracks every once in a while or we'll torture ourselves to death with it. Jesus didn't carry big reference books with Him, and He wasn't a college graduate. He spoke to those poor fishermen and to the sick and the poor and the people who were ostracized and thrown in jail."[51] Day's faith, which was fundamental to her life, was based in the real world of the majestic ocean and the concrete realities of poor people. It is hard to imagine that her kind of faith would turn attention away from rising seas, particularly as those seas began to harm human beings.

## Faith to Face Suffering

For Dorothy Day, the Bible provided little insight into atmospheric condi-tions, but it offered profound instruction on how to treat others. A partic-ularly formative biblical text for her was a discussion of judgment in the twenty-fifth chapter of the Gospel of Matthew, in which Jesus describes God separating the blessed from the wicked. The blessed are told, "I was hungry and you gave me food, I was thirsty and you gave me something to drink, I was a stranger and you welcomed me, I was naked and you gave me clothing, I was sick and you took care of me, I was in prison and you visited me." The wicked, conversely, are those who did not feed, welcome, or visit Jesus. In the Gospel, both groups ask when this happened, unaware that they had fed or failed to feed Jesus. He answers, "Just as you did it to one of the least of these who are members of my family, you did it to me."[52] Day believed in this story at a deep level—she believed that whenever she met a person in need, she was meeting Jesus Christ.

Faith, for Day, required her to pay attention to other human beings and their struggles. She wrote: "It has often seemed to me that most people instinctively protect themselves from being touched too closely by the suf-fering of others. They turn from it, and they make this a habit."[53] She was concerned about how this "habit" of ignoring suffering allowed so many people in New York City and throughout the world to go about their days not doing anything for—indeed, not even noticing—the desperately hungry poor around them. This problem is still very real, and contemporary Catholic Workers continue to help others see and respond to the suffering of the poor.

Anyone who denies the reality of climate change is refusing to pay attention to human suffering. To insist that the oceans are not rising allows one to ignore the sufferings of the people in Bangladesh and Kivalina who are threatened by encroaching waters. To deny that human activity could possibly change the climate is to ignore the suffering of drought-stricken communities in the Sudan, future generations impoverished by a more difficult world, and other species dwindling as ecosystems change. Day's witness suggests that perhaps John Shimkus is not only appealing to the story of Noah to make an argument about sea levels but also is ignoring the deeply troubling realities of this violent world.

Day believed in a God of peace, an almighty being who created the Earth with love and treated it with mercy. But she held this belief in tension with a full awareness that the world is full of violence and pain, that God's creatures suffer and die with no apparent justice. Her faith did not protect her from these facts or explain them away but rather called her to wrestle with a world that would crucify her God and a faith that love could ultimately conquer such evil. Thus, her faith gave her reasons to face the truth as she resisted the violence of her world. And her belief in God helped her to try to bring comfort to those who suffered.

In the years leading up to World War II, when many were denying the realities of anti-Semitism and racism across the world, Day's faith helped her to see these realities of structural violence. She criticized "Christians who affront Christ in the Negro, in the poor Mexican, the Italian, yes and the Jew." God is present in all people, she insisted, and so Christians are called to extend "love and compassion" to everyone, especially those who are marginalized or oppressed.[54] True faith is about noticing the suffering of others and responding to it. I extend this lesson to suggest that, in our own time, Christians who deny or ignore climate change are rejecting "the least of these" and are therefore rejecting Jesus. Christians should see Christ in the poor Bangladeshi driven from her home by flooding, the poor Sudanese refugee who can no longer grow food on his ancestral land, the people of Kivalina whose homes are disappearing, and even polar bears who are losing their habitat and struggling to feed their cubs.

Most privileged people have learned to ignore the suffering of others. We develop deeply ingrained habits that help us block out the pain of those who are not like ourselves. A faith like Day's is one way to overcome these bad habits. Christians should learn to see Jesus, the incarnation of God and the center of all holiness, in every person who suffers from the violence of climate change. Yet this is, of course, not the only religious tool that helps one to see suffering—Buddhist meditations on universal compassion, Islamic laws dictating care for widows and orphans, and Jewish commandments to repair the world similarly seek to move people away from bad habits. This is the power of traditional religions' "ancient dreams." They offer a longstanding commitment to the care of others. One value of religion is that it helps to shape the commitments and habits through which people can learn to see the sufferings of their neighbors, both nearby and around the world.

Of course, people need not be religious to take suffering seriously. As Day's deep respect for her atheist friends demonstrates, many who practice no religion and hold no traditional faith do incredible work on behalf of others. Some climate activists are motivated by faith in the human spirit or in the interconnected systems of nature. Such faith, too, can be a powerful motivation. Day's witness need not be understood to require any particular faith tradition or even traditional faith. Instead, she suggests that one must have faith in something in order to do the hard work of resisting the violence of the world. A foundation in some deep value, a love for something, is vital for developing the courage needed to face the world's suffering and to maintain hope despite the depths of this suffering.

### Faith to Protest

Dorothy Day's witness is evidence of David Loy's argument that "genuine religion" can offer meaningful resistance to structural violence. More specifically, Day, like Loy, believed that religious traditions should stand up against the dominant economic structures and assumptions of our time. Her commitment to voluntary poverty was not just about self-discipline but also about resisting the systems that force others into poverty. For example, selling her newspaper for a penny a copy was a way to reject the profit motive that she saw warping the souls of so many middle- and upper-class people around her. Thus, her faith can help contemporary climate justice activists reflect not only on the place of religion in the movement but also on the place of existing economic systems.

Bron Taylor's interest in dark green religion opens him to the idea that corporations like Sanyo and Disney can be a constructive part of environmental solutions. A profit motive might encourage these corporations to produce and market their products to people who have a spiritual relationship with nature. But, as Loy suggests, such efforts do not stop corporations from seeking quarterly profits, using up resources, and filling their customers with insatiable desires for ever more gadgets.[55] This difference between Taylor and Loy suggests yet another question: Should the climate movement resist capitalism?

The Christian theologian Sallie McFague draws upon the witness of Dorothy Day to say that it should. McFague argues that religion calls for resistance to the dominant economic system in the world today: "The current model of market capitalism has been tried and failed. It is failing in a spectacular fashion," as demonstrated by economic and atmospheric turmoil. The best way to resist is to forgo the trappings of wealth and comfort, to follow Day's example and embrace voluntary poverty. This choice "will cause us to use *all* our considerable assets, at personal, professional, and public levels, to seriously reduce energy use and bring about a new way of being in the world, a way that moves . . . to a wide-open, inclusive view of who we are, a view that has no limits."[56] The structural violence of climate change was caused by well-off people inflicting their desires and wishes upon the world. The solution, McFague argues, is to resist this temptation, to humbly pull back and have less impact on the planet. Those who seek to resist climate change are called to resist the economic system by embracing voluntary poverty.

The logic of this argument is sound. If climate change is disproportionately caused by the rich rather than the poor, then one way to stop contributing to the problem is to stop being rich. This has the added virtue of putting one into community with the victims of climate change, of empathizing with them by sharing their reality.

In some ways, this harkens back to the purification taught by John Woolman. But Day lived in the twentieth century rather than the eighteenth and so was closer to today's economic reality. She was also more interested in political and economic structures than Woolman and so was more strategic about how her voluntary poverty could develop into an organized movement resisting structural violence. A voluntary poverty inspired by Day would involve forming and joining intentional communities committed to resisting climate change, like the farms run by many contemporary Catholic Workers.[57] It would involve using these communities to stage public resistance in ways that would attract notice.[58]

This is a big ask. To forgo all the trappings of financial success is profoundly countercultural. It means not only cutting out problematic behaviors like eating meat and flying on planes but also actually giving up the wealth that provides one the freedom to make such choices. Day's witness asks concerned people to consider that we should not be ruled by what the culture around us considers normal or sensible.

To give up one's wealth in the twenty-first century would be foolish and impractical. But, in Day's view, that does not make it a bad idea. To give a diamond ring away to one woman when hundreds were hungry was not practical. To send the City of New York's interest check back when bills were due was not practical. To fight for the right to vote and never use it was not practical. Yet Day sought to be a witness for an alternative way of living, and that meant rejecting practicality. If business as usual is fundamentally destructive, then one should at least consider a life that refuses to engage in business at all. Because Day believed herself to be utterly dependent upon God, she did not feel dependent on worldly approval or market systems. Her witness suggests that deep resistance to climate change will require a faith—whether Catholic or otherwise—that creates alternatives.

It is entirely possible that this particular form of resistance is a bad idea. Bron Taylor's detection of genuinely dark green religion in profit-driven corporations suggests a reasonable hope that the market might create a sustainable future while people still profit from it. Many others have argued in far more detail that only capitalist markets can organize global efforts efficiently enough to solve the problem of climate change. These arguments should be taken seriously.[59] But Day's witness should also be taken seriously, and those who seek climate justice should consider the possibility that voluntary poverty is an important response to the violence of climate change.

Previous chapters have made it clear that I have not chosen voluntary poverty. I own a computer, fly on planes, live in a single-family house, and enjoy the trappings of middle-class life in the twenty-first-century

United States. But Day's voice helps me to question this lifestyle, asking how much the comforts of my life prevent me from fully resisting the economic structures that cause climate change. Living with these questions, learning from the witness of a radical who chose voluntary poverty, makes me more aware of my choices and more aware of others' suffering. I hope that this makes me a little less likely to become one of the "atheists indeed" who does not resist the violence around me.

Dorothy Day knew that most people, even most who shared her faith, would not devote their lives to voluntary poverty. People who encountered her made many different choices about how to respond. Some committed the rest of their lives to the Catholic Worker, some spent a few months and then moved on, and some never lived with the poor but donated and supported those who did. Day had room for all these people in her movement. "Not all are called," she wrote, "We do what we can." She continued: "If you are a student, study, prepare, in order to give to others, and keep alive in yourself the vision of a new social order."[60] She insisted most strongly not on one way of life but on a vision of a life lived fully out of love for others. Those of us who seek climate justice should think about the ways of life that would express such a vision in the twenty-first century.

## FAITH, LOVE, AND THE CLIMATE

As mentioned above, Dorothy Day told her biographer that the source of her accomplishments in life was the fact that she "wasn't embarrassed to talk about God." The argument of this chapter has been a bit broader than this, but it might be expressed by the idea that no one who seeks climate justice should be embarrassed to talk about faith. This faith may be in a God much like Day's, or it may be in something else entirely. But we should be open to talking about the power of faith to inspire resistance, to help us and our neighbors to see the suffering of others, and to commit to life on Earth.

When Day showed up to picket Ohrbach's department store, she was motivated by a deep faith that drove her to love the striking workers, the police arresting them, and the Church that she believed could unite them. Her faith drove her to love, and her love drove her to protest. Because this protest involved quoting the pope, it disrupted a simplistic narrative that many people had developed—that radicals were atheists calling for chaotic change, while the status quo reflected good and stable Christian values. As a resolute Christian who sided with radicals, Day forced people to question the relationship between their faith, their love, and the violence of the status quo.

A witness of resistance against the violence of climate change will only make sense if it is supported by a belief system. Those who resist the dominant social order with no coherent reasons are easily dismissed as crazy. Those who resist the dominant social order as an expression of a profound faith and a deep love are far harder to ignore. The tradition and calling that empowered Day was her deep Catholic faith. Others are empowered by other faiths—in Allah's mercy, in Buddhist practice, in Krishna's guidance, in Gaia's interconnectedness, in the inventiveness of the human spirit, or in something else. But it takes faith in something to lovingly resist violence. Climate justice activists must be unembarrassed to act on such a faith.

## NOTES

For biographies of Dorothy Day, see especially Miller, *Dorothy Day*; Coles, *Dorothy Day*; and Forest, *All is Grace*. For analyses of Day's thought and importance, see especially O'Connor, *Moral Vision of Dorothy Day*; Thorn, Runkel, and Mountin, *Day and the Catholic Worker Movement*; and Klejment, "Spirituality of Day's Pacifism."

1. Day, *Long Loneliness*, 152.
2. Day, *By Little and By Little*, 87.
3. Forest, *All Is Grace*, 26–27.
4. Day, *Union Square to Rome*, 88.
5. Ibid., 155; also see 131.
6. Ibid., 17.
7. Throughout her life, Day referred to Peter Maurin as "the founder of the Catholic Worker Movement" and gave him a great deal of credit for both the newspaper and the houses of hospitality. There is no doubt that his ideas inspired her and many others. However, Day was far more responsible for the day-to-day organization, the leadership, and the realized ideals of the movement. As Mel Piehl puts it: "Dorothy Day herself promoted the fiction that the Catholic Worker was simply an attempt to realize Peter Maurin's 'Idea.' But it was her common sense and awareness of American social and cultural realities that enabled her

to distinguish between the kind of religious idealism that could inspire a viable social movement in this country and fantastic notions that would merely look ridiculous." Piehl, *Breaking Bread*, 62.

8. Forest, *All Is Grace*, 124.

9. Miller, *Dorothy Day*, 284. A helpful oral history of the movement is given by Troester, *Voices from the Catholic Worker*. The number of Catholic Worker houses shrank during World War II, but it climbed again and, at the time of this writing, there are about two hundred. The anarchist character of the movement makes it difficult to keep such counts reliably, but a list of many current communities can be found at www.catholicworker.org/communities/volunteers.html.

10. Miller, *Dorothy Day*, 255; Roberts, *Day and the Catholic Worker*, 40.

11. Quoted by Forest, *All Is Grace*, 130. Robert Coles writes that this willingness of Catholic Workers to be poor for the sake of the poor is fundamental to the movement. Day and others who chose to be part of the movement become "workers" precisely so that they can develop solidarity with those who are struggling to stay alive, and so the volunteers "merge with those who, in the conventional sense, would be regarded as needing help." Coles, *Dorothy Day*, 111.

12. Day, *On Pilgrimage*, 250.

13. Cornell, Ellsberg, and Forest, *A Penny a Copy*, 25.

14. Ibid., 38. For a compelling account of how Day's pacifism synthesized her early radical beliefs with the Catholic tradition, see Klejment, "Spirituality of Day's Pacifism." For a discussion of the influence her ideas have had on the Roman Catholic peace tradition, see Klejment and Roberts, *American Catholic Pacifism*.

15. Miller, *Dorothy Day*, 377. For more on these changes, see Sicius, "Prophecy Faces Tradition," 66–76.

16. Cornell, Ellsberg, and Forest, *A Penny a Copy*, 107.

17. O'Connor, *Moral Vision of Dorothy Day*, 68.

18. Pope Francis, "Visit to the Joint Session of the United States Congress," September 24, 2015, https://w2.vatican.va/content/francesco/en/speeches/2015/september/documents/papa-francesco_20150924_usa-us-congress.html.

19. Day, *By Little and By Little*, 99. For an insightful comparison of Day with Addams, see Hamington, "Two Leaders, Two Utopias."

20. Dostoyevsky, *The Brothers Karamazov*, book 2, chapter 4.

21. Day, *On Pilgrimage*, 192, 124.

22. Quoted by Riegle, *Dorothy Day*, 81.

23. Day, *On Pilgrimage*, 192.

24. Day, *By Little and By Little*, 242.

25. Day, *Union Square to Rome*, 150.

26. This story is frequently told, but Rosalie Riegle calls it "hard to pin down," as she knows of no one who was present for the exchange. It is possible that the story

is apocryphal, but many who worked with Day believe that it accurately reflects her character. Riegle, *Dorothy Day*, xv.

27. Day, *By Little and By Little*, 294.

28. "An anarchist then as I am now, I have never used the vote that the women won by their demonstrations before the White House during that period." Day, *On Pilgrimage: The Sixties*, 304.

29. Coles, *Dorothy Day*, 97. For Day, this was consistent with a commitment to living with the poor because she believed "efficiency" and "organization" were often coded language used to blame and dismiss the poor for their personal failures rather than attending to institutional forces. When people called her movement impractical, she wrote, "What they are really criticizing is our poverty, the fact that we spend money for food instead of for paint and linoleum. We are crowded as the poor are, with people sleeping in every available corner. We have no separate room for the clothes that come in; they are packed in boxes around the dining-room and hung in one hall closet and in another closet off the dining-room. We are often dirty because so many thousands cross our thresholds. We are dirty ourselves sometimes because we have no hot water or bath, because we have not sufficient clothes for changes,—even because we are so busy with the poor and sick that it is hard to take time to journey to the public baths to wash." Day, *House of Hospitality*, 130.

30. Ibid., 148.

31. Day, *By Little and By Little*, 295.

32. Ibid., 271.

33. Coles, *Dorothy Day*, 29.

34. See especially Gottlieb, *Greener Faith*; and Grim and Tucker, *Ecology and Religion*.

35. Loy, "Religion of the Market," 275–76.

36. Ibid., 283.

37. Ibid., 289. See also Loy, *Money, Sex, War, Karma*; and Stanley, Loy, and Gyurme, *Buddhist Response to the Climate Emergency*.

38. Taylor, *Dark Green Religion*, 13, also see 10. See also Taylor, "From the Ground Up."

39. Taylor, *Dark Green Religion*, 221. For more of Taylor's argument that religion and a sense of the sacred remain important despite arguments for more purely rational and scientific perspectives, see Taylor, "Sacred or Secular Ground."

40. Quoted by Taylor, *Dark Green Religion*, 215.

41. Ibid., 138.

42. In support of this claim, Schweitzer cites a famous essay by Lynn White Jr., "The Historic Roots of Our Ecologic Crisis," which argues that the Judeo-Christian religion justified environmental degradation by teaching people that humanity is not truly a part of nature and that all other creatures were created to serve

human beings. White critiqued Christianity as "the most anthropocentric religion the world has ever seen." However, he went on to argue that Christianity could be reformed and appealed—as the pope would more than fifty years later—to Saint Francis as a model for this reform. This conclusion has received far less attention than his critique, and many environmentalists—like Schweitzer—have since asserted that Christianity and other historic religions are simply too complicit in the current system to possibly fuel a meaningful resistance. White, "Historical Roots of Ecologic Crisis," 1206.

43. Jeff Schweitzer, "Climate Change and Christian Values," June 10, 2009, www .huffingtonpost.com/jeff-schweitzer/climate-change-and-christ_b_198047 .html.

44. Schweitzer and Sciar, *New Moral Code*, 17–19.

45. Jeff Schweitzer, "We Are All Atheists," March 17, 2015, www.huffingtonpost.com /jeff-schweitzer/we-are-all-atheists_b_6890056.html.

46. Day, *House of Hospitality*, 203.

47. Darren Samuelsohn, "John Shimkus Cites Genesis on Climate Change," November 10, 2010, www.politico.com/news/stories/1110/44958.html.

48. Day, *By Little and By Little*, 91.

49. Day, *Union Square to Rome*, 176.

50. This use of Christian faith to more fully invest in the Earth—including earthly tragedies—is perhaps best exemplified by Day's contemporary, the German theologian Deitrich Bonhoeffer. For environmental interpretations of Bonhoeffer's life and work, see especially James B. Martin-Schramm, "Lutheran Theology and the Environment: Bonhoeffer, the Church, and the Climate Question," March 1, 2013, www.elca.org/JLE/Articles/101; and Rasmussen, *Earth Community, Earth Ethics*, chap. 21.

51. Coles, *Dorothy Day*, 70–71.

52. Matthew 25:31–46.

53. Day, *Union Square to Rome*, 8. She also wrote: "Our greatest danger is not our sins but our indifference. We must be in love with God. It is not so much to change what we are doing, but our intention, our motive. It is not sufficient that we refrain from insulting a person; we must love." Day, *On Pilgrimage*, 191.

54. Day, *Union Square to Rome*, 151–52. She went on to emphasize that this was a duty that extended to all humanity: "Catholics believe that man is the temple of the Holy Ghost, that he is made to the image and likeness of God. We believe that of Jew and Gentile. We believe that all men are members or potential members of the Mystical Body of Christ and since there is no time with God, we must so consider each man whether he is atheist, Jew or Christian."

55. Bron Taylor is, of course, fully aware of these critiques. He notes, for example, that many have criticized Disney for the ways it "erodes global cultural diversity,

destroys wildlands to build its parks, promotes consumerism, and celebrates a version of the United States that justifies the deracination of American Indians." He argues that critics "fail to see that some of the company's productions promote progressive politics and have affinities with dark green religion." Taylor, *Dark Green Religion*, 132.

56. McFague, *Blessed Are the Consumers*, 213; also see 77.

57. For an account of life at one rural community inspired by the Catholic Worker and lessons learned from that life about peace and sustainability, see Shanley, *Many Sides of Peace*.

58. Perhaps the most famous such action, inspired by faith but not as far as I know by the Catholic Worker, was Tim DeChristopher's bidding on an oil and gas lease with money he did not have to protest the destruction of land in Southern Utah. DeChristopher served two years in prison but gained considerable attention for his action and remains a leader among environmental activists. See especially Williams, "What Love Looks Like"; and Stephenson, *What We're Fighting for Now*, chap. 6.

59. Kathryn Blanchard and I have attempted to do this in "Prophets Meet Profits: What Christian Ecological Ethics Can Learn from Free Market Environmentalism," *Journal of the Society of Christian Ethics* 33, no. 2 (2014). A well-made argument in favor of free market environmentalism is given by Scruton, *Think Seriously About the Planet*.

60. Day, *By Little and By Little*, 180.

# 6

## Martin Luther King Jr.'s Hope for an Uncertain World

> When our days become dreary with low-hovering clouds of despair, and when our nights become darker than a thousand midnights, let us remember that there is a creative force in this universe, working to pull down the gigantic mountains of evil, a power that is able to make a way out of no way and transform dark yesterdays into bright tomorrows. Let us realize the arc of the moral universe is long, but it bends toward justice.
> —Martin Luther King Jr., *Testament of Hope*

Anyone who has studied the structural violence of climate change will know the "low-hovering clouds of despair" mentioned by Martin Luther King Jr. in this quotation.[1] Humanity's unintentional experiment with Earth's atmosphere is deeply disturbing and profoundly disheartening. The melting shelves of Antarctica and the rising seas of Bangladesh represent disruption and destruction on an epic scale. The shifting borders of forests and the extinction of species represent monumental changes in ecosystem functioning. Homes destroyed by wildfires in California, landslides in Washington, and floods in New Jersey represent the dangers of climate change even for privileged people in the industrialized world. The violence of climate change is multitudinous, but every instance tells a common story: The world is becoming more hostile, more difficult to live in.

Previous chapters have reflected on how to act in response to these troubling facts: by purifying one's own life while still participating in the wider society; by engaging local, regional, and planetary efforts; and by lovingly and faithfully resisting violence. This chapter builds on these ideas but also ponders how to maintain motivation for such action in the face of despair. The violence of climate change raises questions about not only what to do but also why to bother. Action requires motivation, and a moral response to climate change will only be effective if activists learn to feel the pain of the planet without becoming paralyzed by it. Both individually and communally, concerned people must come to terms with what is happening to the Earth, to our corners of it, and to the neighbors with whom we share both.

Many people who engage with the reality of climate change quickly seek refuge in one of two emotional responses: an optimistic expectation that everything will work out or a pessimistic resignation that the world as we know it is doomed. Optimists tend to look to economic and technological developments as promising solutions for every problem posed by climate change. Pessimists tend to emphasize that the atmosphere has already been so fundamentally changed and that industrial societies have been complacent for so long that the only thing to do is mourn and repent for what will be lost. Both these responses are explored in more detail later in this chapter, but here it is most important to note what they have in common: They take away all responsibility to do anything about climate change. An optimist can trust that innovators are hard at work solving the world's problems, and a pessimist can trust that no action will make any difference. Neither one has any reason to act.

This chapter seeks to help climate activists navigate an alternative to these dismissive temptations with a stance of hope. Most simply understood, hope is trust that a better future is possible.[2] In ethical terms, however, hope is more than an attitude; it is a virtue. As such, hope shapes not only how people feel but also who people are and how they react to the world around them; it is not simply a prediction about the future, it is also an orientation that spurs action. Climate justice activists need not only to feel hope but also to be shaped by hope as an alternative to both optimism and pessimism. In response to the structural violence of climate change, hope means believing and behaving and being in a different way. It means living with confidence that human beings can change the future, that people have the power to live differently enough to slow and redress some of the harm to our atmosphere, to other creatures, and to human beings in both the present and future. Hope in a world of climate change means living as though human beings in the industrialized world can cause less harm if we seek to, and those who have been marginalized and oppressed can help lead the way to a better and less violent world.

This chapter explores the virtue of hope in conversation with the words and works of Martin Luther King Jr. King lived out hope in the face of a genuine temptation to despair. He created a sophisticated movement of hopefulness, and he opposed both optimists who believed that justice was inevitable and pessimists who believed it was impossible. His witness models the type of profound and realistic hope that is essential for any response to the violence of climate change.

## STANDING UP FOR RIGHTEOUSNESS

Martin Luther King Jr. was born in 1929 to a middle-class African American family in Atlanta. He described his childhood as filled with love, and he credited the hope that characterized his public life to the secure environment in which he was raised.[3] However, his life was also shaped by the profound injustices of a racist society. He grew up amid "not only segregation but also the oppressive and barbarous acts that grew out of it," bearing witness to police brutality, legalized prejudice, the terrorism of the Ku Klux Klan, and profound economic inequalities.[4] At the age of fourteen years, he was forced to get up from his bus seat by a white man and then had to stand for 90 miles—an injustice that he said "will never leave my memory. It was the angriest I have ever been in my life."[5]

King's work was fueled by that anger—anger at the violence of segregation, racism, and injustice; anger that expanded to include the injustices done to his children, his neighbors, all people of color in the United States, and all oppressed peoples across the world. He devoted his life to resisting this violence.

King was the son and grandson of clergymen, and he attended seminary and then a doctoral program in theology to train as a Baptist minister and intellectual. Bringing the insights of the black Church into conversation with academic discourse, he insisted that Christians are called to a faith that repairs the world's injustices. This is captured well in a 1964 address: "This is the world for which Christ died. We cannot sit idly by and watch it destroyed by a group of insecure and ambitious egotists who can't see beyond their own designs for power."[6] Like Dorothy Day, King dismissed any faith that would ignore Earth for sake of heaven. He preached, "It's all right to talk about long white robes over yonder, but I want a suit and some shoes to wear down here. . . . It's all right to talk about the new Jerusalem. But one day we must begin to talk about the new Chicago, the new Atlanta, the new New York, the new America."[7]

In 1955 King was a relatively new pastor in Montgomery, Alabama, when Rosa Parks was arrested for refusing to give up her bus seat to a white man. It was his commitment to justice that led the twenty-six-year-old King to accept leadership of the subsequent bus boycott. He organized, rallied, and inspired the African American citizens of his city to carpool or walk in protest for an entire year.

Montgomery was King's first public campaign, and it set the tone for the rest of his life by combining thoughtful political strategy with deep moral commitment. The strategic thinking of activists led King and eighty-nine other Montgomerians to willingly serve jail time when they were indicted for civil disobedience. King was following in Gandhi's footsteps, demonstrating that the system punished peaceful African American boycotters but not the white segregationists who inflicted violence upon them. People across the United States began to pay attention, to add energy and resources to the Montgomery protest.

Deep and abiding faith helped King to persevere in this work when he, his wife, and his daughter began receiving death and bomb threats. Late one night at his kitchen table, agonizing over whether the protest was worth the risks, he "experienced the presence of the Divine. . . . It seemed as though I could hear the quiet assurance of an inner voice saying: 'Stand up for righteousness, stand up for truth; and God will be at your side forever.'"[8] He found the courage to struggle for justice in his faith, and he found the meaning of his faith in the struggle for justice.

The Montgomery bus boycott succeeded when the Supreme Court ordered the city to desegregate its public transit system, and King became a figure of national renown. He then led similar campaigns in Albany, Georgia; Birmingham, Alabama; Saint Augustine, Florida; Selma, Alabama; Chicago; and Memphis. His reputation grew internationally, and he became the youngest person ever to receive the Nobel Peace Prize in 1964. As the years went by, his fame grew, but he remained a deeply controversial figure in US society, particularly when he publicly insisted that the violence of racism was inseparably linked to the structural violence of economic inequality and militarism. In 1967 he spoke out against the Vietnam War, saying that he could not lead a nonviolent movement "without having first spoken clearly to the greatest purveyor of violence in the world today—my own government."[9] In 1968 he began to organize a multiracial Poor People's Campaign to educate the public and demand political action for economic justice.

King's work for racial justice, global peace, and economic reform were all ongoing when he was assassinated in Memphis on April 4, 1968. In his final public address the previous night, he once again took courage from his faith and his community, closing his remarks this way: "I just want to do God's will. And He's allowed me to go up the mountain. And I've looked over. And I've seen the promised land. I may not get there with you. But I want you to know tonight, that we, as people, will get to the promised land. So I'm happy tonight. I'm not worried about anything. I'm not fearing any man."[10]

## THE ARC OF THE MORAL UNIVERSE

King's final appeal to "the promised land" is but one example of the virtue of hope in his thinking and his activism. He fought for equality because he hoped that the United States could live up to the lofty democratic rhetoric in its founding documents. He educated people about the interconnection of all things because he hoped that human goodness could overpower fear and spite. He struggled to heal racial divides because he hoped that people could prosper more fully together than divided. And he remained committed to the idea that the world could be changed nonviolently because he hoped and believed that the God of Jesus is always on the side of justice, liberation, and peace.[11]

King's hope was, on one level, based on his understanding of history. He observed that moral progress had been made before, and he predicted that it would continue:

The past is strewn with the ruins of the empires of tyranny, and each is a monument not merely to man's blunders but to his capacity to overcome them. While it is a bitter fact that in America in 1968, I am denied equality solely because I am black, yet I am not a chattel slave. Millions of people have fought thousands of battles to enlarge my freedom; restricted as it still is, progress has been made. This is why I remain an optimist, though I am also a realist, about the barriers before us.[12]

King cited the Emancipation Proclamation, the Supreme Court's striking down separate-but-equal laws, and the decline of colonialism in Africa and Asia as signs of progress, and each one gave him reason to hope.[13]

However, the deepest foundation for King's hope was not human history but faith. He believed that the world is made to favor love and justice, to reward those who build community and defeat those who tear it down. In an early sermon, he put it this way: "Love is the most durable power in the world. . . . He who loves is a participant in the being of God. He who hates does not know God."[14] Nine years later, he made this same point in even more cosmological language: "If there is peace on earth and good will toward men, we must finally believe in the ultimate morality of the universe, and believe that all reality hinges on moral foundations." He continued by exploring the basis of this theological hope in tragic violence: "Christ came to show us the way. Men love darkness rather than the light, and they crucified him, and there on Good Friday on the cross it was still dark, but then Easter came, and Easter is an eternal reminder of the fact that the truth-crushed earth will rise again."[15]

King did not ignore the very real injustices in the world. However, he preserved a hopeful trust that reality is on the side of justice. He inherited this faithful and realistic hope from his Church and his culture, which had preserved faith in God's goodness through the horrors of slavery, the profound injustices of institutionalized racism, and the ongoing oppression of segregation.[16] Giving eloquent voice to this tradition, he insisted that people could preserve hope even as they paid careful attention to the gross injustice done to the poor and to people of color.

King's most common expression of this hope was the phrase with which this chapter began, "the arc of the moral universe is long, but it bends toward justice." He borrowed these words from the nineteenth-century abolitionist Theodore Parker because King understood his stand for justice as a continuation of the same stand made in the previous century.[17] Drawing on this legacy, King emphasized that the world could and would move toward a better future.

Hope helped to justify King's nonviolence. He saw violence as a departure from the moral order, a push against the universe's fundamental direction. Nonviolence, conversely, is participation in the community of creation, cooperation with the interdependence and justice that characterize reality at its core. To hurt others is to fight against the universe. To refuse to hurt others, even to accept suffering in the service of justice, is to bend along with the universe.[18]

In his most famous public oration, the 1963 "I Have a Dream" speech in Washington, King emphasized that genuine hope must stir action: "We can never be satisfied as long as our children are stripped of their selfhood and robbed of their dignity by signs stating 'for whites only.' We cannot be satisfied as long as a Negro in Mississippi cannot vote and a Negro in New York believes he has nothing for which to vote. No, we are not satisfied, and we will not be satisfied until justice rolls down like waters and righteousness like a mighty stream."[19]

As King's thought continued to develop in the last years of his life, he increasingly emphasized the importance of dissatisfaction, calling his audiences to protest against any system or person who refused to conform to the uplifting shape of the universe. In 1967 he lamented that, shortly after his famous speech "about that dream, I started seeing it turn into a nightmare," when white supremacists bombed a church and killed four young girls, when thousands died of preventable illnesses in the richest nation in the world, when tens of thousands were killed overseas as his country expended ever more resources on war. "Yes," he preached, "I am personally the victim of deferred dreams, of blasted hopes, but in spite of that I close today by saying I still have a dream, because, you know, you can't give up in life."[20]

Still, the reality of injustice tested King's hope. The anger he learned at the age of fourteen intensified as he witnessed the depth of the violence brought against his movement and the oppressed peoples of the world. Early in his public career, he had emphasized that he wanted African Americans to participate fully in existing structures—indeed, as late as 1966, he was reassuring moderates that his movement sought to protect "the American economy, the housing market, the educational system and the social opportunities" so that all people could enjoy them.[21] However, by the end of his life, he sounded far more countercultural. He wrote, "White America must recognize that justice for black people cannot be achieved without radical changes in the structure of our society." He insisted, "For the evils of racism, poverty, and militarism to die, a new set of values must be born. Our economy must become more person-centered than property- and profit-centered. Our government must depend more on its moral power than its military power."[22]

By 1968 King's hope had become revolutionary. He now insisted that to be on the side of the universe meant not only ending segregation but also dealing with the structural roots of racism, poverty, and militarism. Vincent Harding, an activist and scholar who wrote speeches for King, argues that this makes King an "inconvenient hero," who should inspire not a holiday of polite remembrance and platitudes but deep and continuing examination of how our nation and our world remain in the grip of violence, inequality, and prejudice.[23]

Despite the increasingly revolutionary nature of his witness, King sustained his core commitment to hope. He never abandoned his belief that the universe is on the side of justice, and he never stopped living out the truth of this idea. This virtuous hope, and King's ability to sustain it in the face of profound violence, can be a vital inspiration for anyone concerned about climate justice in the twenty-first century. Although his hope is profoundly theological, those who learn from it need not agree with every aspect of his faith. Those coming from different religious traditions or no tradition may not share his assertion that God made a universe that bends toward justice, but there are other reasons to live as though the universe can bend toward justice and that a better future is possible. The power of King's witness is that he shows how to make such a hope realistic and practical in a violent world.[24]

## THE SYNTHESIS OF HOPE

Part of Martin Luther King's genius was his ability to hold ideas in tension, to seek a synthesis by balancing concepts that initially seem incompatible.[25] Thus, the vision of hope he models is best understood as a series of balances. This section explores two such balances that are explicit in his writings: between human goodness and human limitations and between militant demands and moderate practicality. The next section then returns to the central concern of this chapter: how to find a balance between despair at the inevitability of climate change and optimism about human ingenuity.

### Between Human Virtue and Vice

King's hope balances faith in human action with a realistic skepticism about human nature. Drawing on the theological debates of his time, he defined one extreme as liberalism, which he understood as a belief that human reason can solve all problems. At the other extreme was neoorthodoxy, defined by the belief that sin is the dominant characteristic of humanity. Balancing these two allowed him to expect a great deal from people while understanding their limitations.[26]

King believed that the universe bends toward justice but that culture progresses in this direction only when people push it. Only human action can make positive change in human systems. Thus, King expected immediate and substantial action from his audiences; he expected them to be good enough to fight for justice. Against those who asked the civil rights movement to be more patient, more tolerant of injustice, he responded that reality would not allow it: "The shape of the world will not permit us the luxury of gradualism and procrastination. Not only is it immoral; it will not work."[27]

However, King was also aware of the limits of human goodness, of very real temptations to resist the arc of the universe. When asked in 1965 about what mistakes he had made, he lamented a naïveté about human nature:

The most pervasive mistake I have made was in believing that because our cause was just, we could be sure that the white ministers of the South, once their Christian consciences were challenged, would rise to our aid. I felt that white ministers would take our cause to the white power structures. I ended up, of course, chastened and disillusioned. As our movements unfolded, and direct appeals were made to white ministers, most folded their hands—and some even took stands *against* us. (emphasis in the original)[28]

Human beings could and would sometimes fail to do good work. King learned this lesson not only from his opponents but also from his colleagues, and so he stressed that nonviolence did not make people perfect. Everyone, even a nonviolent activist, wrestles with moral limitations and sometimes fails.[29]

Sustainable hope requires trust in human beings balanced with an awareness of human limitations. It requires an awareness of what King called "a strange dichotomy of disturbing dualisms within human nature" between a capacity for good and a tendency for evil.[30] Recognizing this dichotomy means knowing that even the opponents lined up against the climate movement are capable of goodness. This means that part of our task is to help them to be good. It also means that the noblest advocates of climate justice are imperfect and will at times fail to live up to their own standards. A hopeful movement for climate justice needs to be aware of this dichotomy of disturbing dualisms and will thrive only if it remains attentive to all sides of human nature.

## Between Militancy and Moderation

King's hope also balances fierce urgency about the troubles of the world with patient awareness that change cannot happen without a deliberate process. In his movement, he often framed this as a sensible middle ground between the extremism of black militants, who demanded immediate change and justified the use of violence to achieve it, and white moderates, who asked for more time to make change and justified the continuation of violence while history unfolded. King argued that these extremists and moderates were not as far apart as they believed themselves to be; "there is a striking parallel—they accomplish nothing; for they do not reach the people who have a crying need to be free."[31] The way to make real change, to respond constructively to the world's violence, is to synthesize the two.

Extremism has its place. King celebrated how the civil rights movement "demonstrated great militancy" by persistently demanding justice in
a nonviolent way.[32] However, he rejected militancy that justifies hurting or
killing others, separating people into categories that allow "us" to attack
"them." A militant hope must stand up for what is right while continuing
to take seriously the diversity and complexity of social problems. Thus,
concerned people today have a responsibility to militantly oppose the violence of climate change. But this need not mean advocating a single or
simplistic answer, it need not mean vilifying all who contribute to atmospheric change, and it need not mean using violence. Climate change has
been a slowly unfolding process, and resolving it will be neither immediate
nor straightforward. Militancy must be balanced with patience.

Though King argued against violent militants who demanded immediate change, he was far more disappointed by complacent moderates who
were too patient with injustice. While imprisoned for civil disobedience
in Birmingham, he received a letter from local clergy that called his protests "unwise and untimely" and suggested that the civil rights movement
should patiently wait for the resolution of court proceedings rather than
protesting in the streets. He responded with one of his most eloquent
pieces of writing, the "Letter from Birmingham Jail." He insists, "We will
have to repent in this generation not merely for the hateful words and
actions of the bad people but for the appalling silence of the good people." Against the appeal to patience, he argues, "Human progress never
rolls in on wheels of inevitability; it comes through the tireless efforts of
men willing to be co-workers with God, and without this hard work, time
itself becomes an ally of the forces of social stagnation."[33] He published
this letter in a book titled *Why We Can't Wait*, yet again emphasizing that
moderate patience is intolerable in the face of injustice.

People concerned about climate change face a similar challenge and
must convince not only those who deny the problem but also those who
agree that climate change is real but do nothing about it. King's message
of urgency, his call to act on the side of the universe, is a vital example
for the twenty-first-century climate justice movement. Nevertheless, there
remains a place for moderation in this movement. King defined proper
moderation as a "virtue," which requires "moving through this tense
period of transition with wide restraint, calm reasonableness, yet militant
action."[34] Moderate action on climate change does not mean giving up

militant demands, but it does mean reasonably assessing the complexity and wickedness of the problem. It means restraining oneself from unhelpful stridency and extremism while nevertheless insisting on justice in a violent world.

King's balance between militancy and moderation teaches concerned people in the twenty-first century to live out our hope by demanding justice and action but never to do so with the narrow vision or certainty that leads to a closed mind or a clenched fist.[35]

## BETWEEN OPTIMISM AND DESPAIR

Perhaps the most difficult balance to strike when thinking about climate change is the one discussed at the beginning of this chapter—between despair at the irreversible damage that has been done to the world on one hand and an optimistic faith that there could soon be solutions to all climate problems on the other. King does not offer specific guidance here; but the example of his hope can offer inspiration for the climate movement. First, I explore each temptation and then consider how King's thought might suggest a synthesis.

### The Temptation to Optimism

As an example of optimism, consider the work of the Danish statistician Bjørn Lomborg, who became famous as "the skeptical environmentalist" when he released a book with this title in 1998. Ten years later, he published a follow-up exclusively about climate change, provocatively entitled *Cool It.* He admits that we face a problem but emphasizes that this "is emphatically not the end of the world." The sea level may be rising, but people managed the sea level rises of the twentieth century and so will doubtless manage continued rising in the twenty-first. He accepts the need for governmental programs responding to climate change, but he emphasizes that such programs must not restrict or limit the use of fossil fuels because the benefits of burning such fuels far outweighs the costs.[36]

Lomborg's approach to climate change is informed by a profound confidence that human beings in a healthy economy can invent solutions to whatever problems they face. Trusting innovation, he believes that climate change will be solved by technology. Properly incentivized people and corporations will create cleaner and better ways to produce energy and food. He encourages environmentalists not to treat climate change as desperately urgent or to dwell on gloomy predictions because such rhetoric causes people to panic rather than carefully weigh the costs and benefits of their actions.

Lomborg worries that most proposed responses to climate change—reducing consumption, limiting fossil fuel use, rolling back globalization—will make the world poorer, and he suggests that the global economy would lose $40 trillion by 2100 if it followed popular political proposals.[37] A less prosperous human race would be less equipped to deal with climate change and any other challenges that arise. People's primary responsibility to the future, Lomborg argues, is to ensure that they become as wealthy as possible. As long as future generations have financial resources, he optimistically insists, they will be better off than their ancestors.

To support his argument, Lomborg offers evidence of the many problems that have already been solved technologically. He writes that human beings "are actually leaving the world a better place than when we got it." As technology and civilizations have developed, human well-being "has vastly improved in every significant measurable field," and "it is likely to continue to do so."[38] People frustrated by the challenges of travel invented cars and jet airplanes; people frustrated by the challenges of communication invented the telephone and the internet. People frustrated by polio and the flu invented vaccines. And so people frustrated by climate change will invent ways to capture carbon emissions while predicting or controlling unpleasant weather patterns. Technology has already solved problems, making humanity healthier and the world better. Lomborg finds no reason to think this trend will not continue.

Lomborg is an extreme example, but he is not the only climate optimist. Indeed, one of the most famous climate change activists in the world, former vice president Al Gore, wrote in 2014 that "the forward journey for human civilization will be difficult and dangerous, but it is now clear that we will ultimately prevail." Gore optimistically cites rapid progress in solar energy production, in economic strategies discouraging investment in fossil fuels, and in political discourse. He believes that "we are finally putting ourselves on a path to solve the climate crisis."[39]

This view of humanity's future is an understandable response to the violence of climate change. Innovations in renewable energy may some-day allow industrial civilization to continue without the fossil fuels that today spoil the atmosphere. A concerted and careful effort of technolog-ical development could potentially rebalance the climate, counteracting the unintentional changes in the atmosphere with intentional engineering. But acknowledging these possibilities is very different from trusting that they will be realized. Seeing what is possible is an essential part of hope. Trusting that good things will happen, by contrast, is an optimism that promises a better future rather than encouraging people to work for it.

Optimism is also dangerous when it turns attention toward the future without a careful look at the present. When trust in technology and eco-nomics leads to lofty claims about how new energy sources and private businesses might one day solve problems, it ignores the people, creatures, and ecosystems already suffering from the violently changing climate. With seas already rising, with droughts already increasing, with food already scarce, climate change is a real and present problem. To answer this fact with claims about what will someday be possible is insufficient; it ignores the suffering of our neighbors and the wickedness of today's problems.

The hope modeled by Martin Luther King Jr. is opposed to easy opti-mism. King made confident predictions about the arc of the moral uni-verse, but he insisted that this arc was long and that concrete change could only come from the diligent, strategic, and concerted efforts of people who resist violence. His hope preserves the possibility of a better future in light of the present reality, while optimism promises such a future without attention to contemporary complexities. Hope inspires modesty, remind-ing us of the uncertainty of what is to come; optimism offers a self-inflating guarantee that humanity will get the future it wants. Hope requires caution that the path ahead may twist and turn; optimism boldly claims to know exactly where humanity is headed.

## The Temptation to Despair

At the opposite extreme from optimism is despair—taking the violence of climate change so seriously that one surrenders to the inevitability of its worst possibilities. The longtime environmental activist Paul Kingsnorth captures this sentiment well: "Whenever I hear the word 'hope' these days, I reach for my whiskey bottle. . . . It seems to be such a futile thing. What does it mean? Why are we reduced to something so desperate? Surely we only hope when we are powerless?"[40] Honest attention to the world's condition, he argues, leads not to hope or optimism but to an acceptance that the twenty-first century is the "age of ecocide."

In a manifesto cowritten with Dougald Hine, Kingsnorth asserts that climate change "brings home at last our ultimate powerlessness," clarifying "the head-on crash between civilization and 'nature.'" Hine and Kingsnorth propose that industrial human civilization has proven itself destructive and unrealistic, incapable of coexisting within the patterns and structures of the natural world. So they call for a movement of "uncivilisation," an artistic project focused on observing and participating in the nonhuman world rather than seeking to save human cultures.[41]

A more systematic argument for despair comes from the ethicist Clive Hamilton, whose book *Requiem for a Species* is devoted to articulating the grim reality of climate change. His core claim is that "we can no longer prevent global warming that will this century bring about a radically transformed world that is much more hostile to the survival and flourishing of all life."[42] Given this truth, he argues, environmentalists make a mistake when they cultivate hope. Telling people that a positive future is possible if enough of them buy the right light bulbs or lobby their legislators simply feeds delusions. Nothing anyone does today will "solve" climate change, and so any discussion of solutions is really "a means of disengaging from reality." Instead, Hamilton argues, those who understand the reality of climate change should be allowed to "enter a phase of desolation and hopelessness—in short, to grieve."[43]

Hamilton and Kingsnorth mourn the loss of a climate that existed beyond and above human influence, of a relatively consistent atmosphere that characterized human history until the twentieth century. They lament that "nature," in the sense of systems untouched by human hands, no longer exists.[44] Mourning is a sensible response to these facts, and this claim that the world is becoming more hostile to the thriving of life is well supported.[45] Climate change is violence.

Furthermore, even the dourest environmentalists argue that such mourning is not a virtue for its own sake but should be a precursor to action. Hamilton insists that grief is simply one step, followed by acceptance and protest; mourning is necessary for "acting, and acting ethically."[46] In their manifesto for "uncivilisation," Kingsnorth and Hine conclude with a call to "the paths which lead to the unknown world ahead of us." None of these authors advises concerned people to simply mourn. Instead, they see despair as a realistic response necessary for modest action.

However, these authors move dangerously close to despair when they discourage hope. It is dangerous to give up on the belief that positive forces can shape the future. Although industrial civilization has caused great harm to both human and nonhuman lives, it is a step too far to give up entirely on all its accomplishments. It is appropriate to look squarely at the irreversible and destructive effects of climate change, but this must be balanced against the danger of giving up in desolation and hopelessness.

The virtue of hope resists despair. It finds a way to believe that a better future is possible in order to roll up its sleeves and get to work. Hope acknowledges the challenges ahead, while despair capitulates to them. Hope laments what has been, is being, and will be lost; it then attempts to respond constructively. Hope insists that the future is unknown and demands that people work to make it better, while despair claims certitude about what will happen next and surrenders the power to change it.

## Hope in a Changing Climate

Despair and optimism both offer clarity about what the future will look like. Optimism means deciding that the world as we know it will be just fine; despair means deciding that the world as we know it is over. Hope, by contrast, begins with uncertainty, with awareness that the future is undetermined. Hopeful people live and act with trust that good things are not guaranteed but are possible if they work toward them. The movement for climate justice should be hopeful, avoiding either form of certainty by denying optimistic trust in technology on one hand and despairing resignation to uncivilization on the other hand.

Still, it is worth conceding that there is something of value in both optimism and pessimism. Martin Luther King called himself "optimistic" at times, and he spun narratives of progress for justice, racial equality, and other dreams of the future. This was part of his rhetorical strategy, and it makes sense that some rhetoric about climate justice today would similarly talk about the steady progress of human morality and technology that could lead to a positive future. When one thinks of the diseases that have been cured thanks to medical innovations, of the global communities that have been created via digital media, and of the massive increases in food stocks that have been grown using agricultural technologies, it is understandable that some people believe in innovative solutions to climate change. It is reasonable for environmentalists like Bjørn Lomborg to insist that the solution to climate change is not social change or moral hand wringing but instead wise investment in economic and technological innovation.

But the temptation to despair is also reasonable. When King's family was threatened with violence in Montgomery, he stayed up all night at his kitchen table wrestling with hopelessness. When four innocent girls were killed in a church bombing, he mourned profoundly, as did all people of goodwill. Perhaps this legacy of mourning can help contemporary climate activists better accept that there is no future without the violence of climate change, no present that does not involve an inundated Kivalina and a desertified Sudan. Paul Kingsnorth and Clive Hamilton are right that the structures of industrial human activity are irredeemably wrapped up in global catastrophe.

If there is a reasonable case to be made for both optimism and despair, then the most sensible way forward is a synthesis of the two. King's witness models such a synthesis because the civil rights movement of which he was part held present violence and future possibilities in creative tension. African American communities maintained hope—even after being torn from their homelands, dragged across the sea on deadly slave ships, owned as property, denied the vote in a nation that claimed to be democratic, excluded from economic wealth and political participation, and systematically exploited and oppressed. King, fully aware of all this violence, stood in front of the Nobel Peace Prize Committee and admitted to "an abiding faith in America and an audacious faith in the future of mankind. I refuse to accept the idea that the 'isness' of man's present nature makes him morally incapable of reaching up for the eternal 'oughtness' that forever confronts him."[47] Calling on the legacy of slave resistance, the Underground Railroad, protest groups, black pride, and the abiding faith of his Church, he lived out the virtue of hope.

Climate activists in the twenty-first century must similarly face the violence of the present and draw on the resources of the past to look forward to a better future. This means learning from King and the civil rights movement, from the other witnesses discussed in this book, and from all others throughout history who have dissented from overly simplistic answers but still made the world better. To resist climate change does not require blind faith that the atmosphere can be returned to some preindustrial state nor an optimistic claim that the violence of global warming will be simply undone. It requires, instead, hope that the world can become better, more just, and healthier than it is today. It requires hope that the future will be better than the present if people work toward the good. And it requires hope that the arc of the moral universe can bend toward justice.

## THE TEST OF CLIMATE ENGINEERING

This chapter has focused on how climate justice activists should nurture hope in the face of "low-hovering clouds of despair," without resorting to simplistic dreams of a bright and sunny road ahead. As King demonstrates, genuine hope is powerful because it can lead to action in personal, political, and global contexts. Therefore, it is appropriate to conclude by considering a more concrete question about how those concerned with climate justice should respond to climate engineering.

The most common definition of climate engineering is "the deliberate large-scale manipulation of the planetary environment to counteract anthropogenic climate change."[48] Climate engineering is an attempt to counterbalance climate change with larger-scale, intentional interventions in the global climate. In contrast to traditional responses to climate change that propose emitting fewer greenhouse gases, climate engineers turn their attention to technological proposals that would sequester such gases, remove them from the air, and/or find other ways to work against their climatic effects.[49]

These proposals can be divided into two categories. First is solar radiation management, which attempts to reflect more of the sun's heat back into space. One such proposal imagines injecting sulfuric acid into the atmosphere, where it would form artificial clouds that would reflect sunlight and thus lower the planet's average temperature.[50] The second category is carbon dioxide removal, which would seek to prevent climatic changes before they happen by removing human emissions from the atmosphere. Attempting to demonstrate the viability of this approach, the entrepreneur Russ George dumped 100 tons of iron sulfate into the Pacific Ocean in 2012 to show that it could fertilize algae blooms, which would in turn absorb large amounts of carbon dioxide.[51]

Most discussions of climate engineering focus on scientific questions. The most common topics concern what is physically possible: Could engineers properly calculate an appropriate artificial cloud cover, and where would the raw materials for this project come from? What impact would algal blooms have on ocean ecosystems beyond the absorption of more carbon dioxide?[52] Other common questions turn to political possibilities: Who would approve, monitor, and pay for geoengineering efforts? Given the limitations of global governance, how could such proposals be adjudicated as legal or illegal?[53] One's answers to these questions tend to determine one's positions on climate engineering.

Bjørn Lomborg, the optimistic author of *The Skeptical Environmentalist* discussed above, advocates climate engineering. He supports research into solar radiation management, identifying this as precisely the kind of technology that will justify continued fossil fuel use and continued wealth creation, and give people in the future the tools they need to thrive.[54] He is confident that scientific challenges can be resolved by good engineering and that efficient governments using market forces to direct climate engineering can manage the global temperature. Such confidence in climate engineering requires faith in human ingenuity, an optimistic belief that technology can create helpful responses to complex problems.

Of course, climate engineering also has strident critics who worry that such optimism is unfounded. Clive Hamilton, the pessimistic author of *Requiem for a Species* discussed above, understands climate engineering proposals as more false hopes designed to delay the despair that must inevitably come from a changing climate. He suggests that human beings should have learned by now that they cannot control the atmosphere, that the natural world is too complicated to be simplistically managed by human science and technology. He also worries that any attempt to engineer the climate will simply exacerbate existing political disparities, continuing the hubristic, oppressive patterns that led to climate change in the first place. So he despairs that any human intervention in the world could ever heal the planet or damaged human communities.[55]

King's hope is powerful because it synthesizes disparate ideas. Because he was utterly committed to nonviolence, he sought to learn from everyone on all sides, and he looked for ways to bring extremes into conversation to find a path forward. What would such a hope look like in response to climate engineering? How might one synthesize the stark dismissal of critics with the bold claims of advocates? Neither blind faith in climate engineering nor complete despair that it could possibly work is a hopeful response. Neither unquestioning belief that people can improve the world nor absolute certainty that they will inevitably make it worse reflects the virtue of hope.

A synthesis is possible if one begins not with the science or politics of climate engineering but with human nature. The most important questions about climate engineering concern not the relevant natural systems or political institutions but the human beings making and affected by the decisions of whether and how to engineer the climate. King titled his final book *Where Do We Go from Here—Chaos or Community?* This book draws on his understanding of violent racial realities but finds ways to develop hope about the future of racial relations as well as the future of global community, economic structures, and technology. On the latter subject, King celebrates the "wonders that science has wrought in our lives," but he is concerned about a "lag" of moral progress that had not caught up with scientific progress: "When scientific power outruns moral power, we end up with guided missiles and misguided men. When we foolishly minimize the internal of our lives and maximize the external, we sign the warrant for our own day of doom."[56]

Science and engineering are capable of wondrous achievements, and it is entirely possible that climate engineering could reduce the violence of climate change. But science and engineering are also capable of unleashing terrible and uncontrolled power, and so it is possible that climate engineering could exacerbate the unjust degradation of the contemporary world. The difference will be made by the human beings who create technology and the social and political systems within which this technology exists. Climate engineering could be a positive force because human beings are capable of good work; but it could also be a disaster because human beings are capable of great evil.

A hopeful response to climate engineering will ask how the scientific progress required to create such technologies could be matched by the moral progress necessary to make wise decisions about whether and how to use it. Wealthy people in the industrialized world cannot be trusted to manage the climate until we have demonstrated that we can better manage the desires, the carelessness, and the hubris that created the problem of climate change in the first place. So my own answer to the question of whether the climate should be intentionally engineered is neither yes nor no; it is not yet.[57]

Inspired by King, I have hope that people like me will one day be mature enough to repair the world we have degraded, to manipulate the atmosphere and make it less hostile to thriving life. I do not share Hamilton's despair about the world's future or about human nature. However, also inspired by King, I see the ways people like me have proven ourselves currently unqualified to manage the climate. We have carelessly degraded the global atmosphere, creating violence. So I believe Hamilton is too militant in his dismissal of human ingenuity, but Lomborg is too moderate in his confidence about that ingenuity.

Regardless of humanity's technological capacity to undertake climate engineering, we are not ready for it in the early twenty-first century. Learning from King about the limitations of human beings, I do not believe that the privileged people who continue to carelessly spew climate-changing gases can be trusted to make intentional changes to the atmosphere. This would be like asking segregationists to design new political systems in the hopes that they would become more inclusive. The privileged people in the developed world who have inherited and exacerbated the problem of climate change are not yet up to the task of reversing it. Our work, instead, must be to reform ourselves and our behavior. We must combine the "wide restraint," "calm reasonableness," and "militant action" that King advocated. This will help us to first stop degrading the climate, before we turn to the project of repairing the atmosphere.

To follow King's hope is to believe that there is a power in the universe seeking to "pull down the gigantic mountains of evil, and "to make a way out of no way and transform dark yesterdays into bright tomorrows."[58] This belief includes the possibility of a better future, where climate change is less violent, less destructive, less troubling. But King teaches that human beings must change themselves before pushing toward this future. The question for concerned people today is not how to shape the world. We are not ready for that. Instead, we should first ask how we can better conform ourselves to the hopeful bend of the universe.

## NOTES

For biographies and analyses of King, see especially Branch, *Parting the Waters*; Baldwin, *Balm in Gilead*; Cone, *Martin & Malcolm & America*; Baldwin, *Make the Wounded Whole*; Harding, *Martin Luther King*; Branch, *Pillar of Fire*; Burns, *To the Mountaintop*; Branch, *At Canaan's Edge*; Sitkoff, *King*; and Burrow, *Extremist for Love*. In this chapter I have drawn heavily on these works and on King's published books, sermons, speeches, and interviews. Some scholars suggest that his ideas are more fully expressed in his unpublished work, in part because a number of his public works were ghost written; see Garrow, "Intellectual Development of King"; and Cone, *Martin & Malcolm & America*. However, I am convinced by historian Lewis Baldwin that the discrepancies between published and unpublished works are not pivotal and that the writings published under King's name express his beliefs and ideas accurately; see Baldwin, *Balm in Gilead*, 13.

1. King, *Testament of Hope*, 252.
2. For a more detailed analysis of hope as a virtue and resource for ecological ethics, see Blanchard and O'Brien, *Introduction to Christian Environmentalism*. Some of the discussion in this section is indebted to chapters 6 and 7 of that book.
3. In a spiritual autobiography assignment for seminary, King wrote: "It is quite easy for me to think of a God of love mainly because I grew up in a family where love was central and where lovely relationships were ever present. It is quite easy for me to think of the universe as basically friendly mainly because of my uplifting hereditary and environmental circumstances. It is quite easy for me to lean more toward optimism than pessimism about human nature mainly because of my childhood experiences." King, *Autobiography*, 2.
4. King, *Stride toward Freedom*, 90.
5. King, *Testament of Hope*, 343.
6. King, *"In Single Garment of Destiny,"* 21.
7. Quoted by Burrow, *Extremist for Love*, 90–91.
8. King, *Stride toward Freedom*, 134–35. For a discussion of how central this experience was to the next twelve years of King's activism, see Garrow, "Spirit of Leadership."
9. King, *Testament of Hope*, 233.
10. Ibid., 263.

11. David Chappel titled his analysis of prophetic religion in the civil rights move-
    ment "A Stone of Hope," drawing the phrase from King's 'I Have a Dream'
    speech about his people hewing "a stone of hope" from "a mountain of despair"
    to exemplify the core belief of the movement. Chappell, *A Stone of Hope*, 1. See
    also Dawson, "Concept of Hope"; Franklin, "An Ethic of Hope"; and Burrow,
    *Extremist for Love*, 245–51.

12. King, *Testament of Hope*, 314. Note that King's self-identification as an "optimist"
    here is more compatible with what I am referring to as hope. As discussed later in
    the chapter, optimism is too often complacent, and King was never complacent.

13. King, *Strength to Love*, 80–82.

14. King, *Testament of Hope*, 11.

15. King, *Trumpet of Conscience*, 75.

16. "King understood how paradoxical notions of pessimism and optimism, of apa-
    thy and hope, of pain and affirmation had always existed side by side in black
    art and thought. In the many versions of his sermon 'A Knock at Midnight,' he
    often mentioned how his slave ancestors combined heartache and hope in their
    songs." Baldwin, *Make the Wounded Whole*, 64. See also Watley, *Roots of Resis-
    tance*, 21; and Cone, *Cross and Lynching Tree*.

17. Rufus Burrow suggests that King likely did not take the phrase directly from
    Parker but from others who quoted him; Burrow, *God and Human Dignity*, 189.
    However, Parker's original sermon is a powerful piece of rhetoric that resonates
    with King's ideas. Consider: "Look at the facts of the world. You see a continual
    and progressive triumph of the right. I do not pretend to understand the moral
    universe; the arc is a long one, my eye reaches but little ways; I cannot calculate
    the curve and complete the figure by the experience of sight; I can divine it
    by conscience. And from what I see I am sure it bends towards justice. Things
    refuse to be mismanaged long. Jefferson trembled when he thought of slavery
    and remembered that God is just. Ere long all America will tremble"; Parker,
    *Collected Works*, 48.

18. "A sixth basic fact about nonviolent resistance is that it is based on the con-
    viction that the universe is on the side of justice. Consequently, the belief in
    nonviolence has deep faith in the future. This faith is another reason why the
    nonviolent resister can accept suffering without retaliation. For he knows that
    in his struggle for justice he has cosmic companionship. It is true that there are
    devout believers in nonviolence who find it difficult to believe in a personal
    God. But even these persons believe in the existence of some creative force that
    works for universal wholeness. Whether we call it an unconscious process, an
    impersonal Brahman, or a Personal Being of matchless power and infinite love,
    there is a creative force in this universe that works to bring the disconnected
    aspects of reality into a harmonious whole." King, *Stride toward Freedom*, 106–7.

19. King, *Testament of Hope*, 218–19.

20. King, *Trumpet of Conscience*, 76.

21. King, *Testament of Hope*, 58.

22. Ibid., 314; King, *Where Do We Go from Here*, 133.

23. Harding wrote in 1996: "Perhaps the memory of Martin King needs to be broken free from all official attempts to manage, market, and domesticate him. At this moment near the closing of his century, we need a truly free and inconvenient hero, one who may help us to explore new dimensions of our freedom, not simply as a private agenda, but to follow his unmanageable style of seeking and using freedom to serve the needs of the most vulnerable, the most unfree among us." Harding, *Martin Luther King*, x. See also Cone, *Martin & Malcolm & America*; Moses, *Revolution of Conscience*; and Hayes, "Hope and Disappointment."

24. The emphasis of this chapter is on hope as an alternative to optimism and despair, which are key temptations in the climate movement as I understand it. Were King brought into a conversation about climate denial—resisting the idea that climate change is a moral problem at all—more could be said about courageous hope as an alternative to fear. Denial seems to me a product of a political strategy that plays on anxieties and uncertainties about the future to drive attention away from scientific evidence and the human experience of our current planetary reality. So, a movement against climate denial would need to develop a constructive response to fear. King would be a vital source for such a response.

25. For example, in response to economic inequality, he argued that capitalism and communism were both limited, calling for a synthesis between them. See King, *Testament of Hope*, 250. King frequently put this in explicitly Hegelian terms. As Rufus Burrow writes, King "was enamored with Hegel's dialectical method of thesis, antithesis, and synthesis as the best means to truth, and was fascinated with the Hegelian idea that growth comes through suffering and struggle." Burrow, *Extremist for Love*, 77.

26. He wrote: "An adequate understanding of man is found neither in the thesis of liberalism nor in the antithesis of neo-orthodoxy, but in a synthesis which reconciles the truths of both." King, *Testament of Hope*, 36.

27. King, *Why We Can't Wait*, 129.

28. King, *Testament of Hope*, 344–45.

29. See especially King, *Stride toward Freedom*, 99. Harvard Sitkoff interestingly argues in *King: Pilgrimage to the Mountaintop* that King's profound awareness of human limitations was informed by a deep personal discomfort with his own adultery.

30. King, *Testament of Hope*, 48.

31. King, *Why We Can't Wait*, 42.

32. King, *Testament of Hope*, 661.

33. King, *Why We Can't Wait*, 86. The letter to which King was responding can be found at https://swap.stanford.edu/20141218223221/http://mlk-kpp01.stanford.edu/kingweb/popular_requests/frequentdocs/clergy.pdf.

34. King, *Testament of Hope*, 661.

35. King wrote: "The nonviolent approach provides an answer to the long-debated question of gradualism versus immediacy. On the one hand it prevents one from falling into the sort of patience which is an excuse for do-nothingism and escapism, ending up in standstillism. On the other hand it saves one from the irresponsible words which estrange without reconciling and the hasty judgment which is blind to the necessities of social progress. It recognizes the need for moving toward the goal of justice with wise restraint and calm reasonableness. But it also recognizes the immorality of slowing up in the move toward justice and capitulating to the guardians of an unjust status quo. It recognizes that social change cannot come overnight. But it causes one to work as if it were a possibility the next morning." King, *Stride toward Freedom*, 221.

36. Lomborg, *Cool It*, 149, 155.

37. "Conclusion," in *Smart Solutions*, by Lomborg, 395.

38. Lomborg, *Skeptical Environmentalist*, 351.

39. Albert Gore, "The Turning Point: New Hope for the Climate," June 18, 2014, www.rollingstone.com/politics/news/the-turning-point-new-hope-for-the-climate-20140618.

40. Daniel Smith, "It's the End of the World as We Know it, and He Feels Fine," *New York Times*, April 17, 2014.

41. Paul Kingsnorth and Dougald Hine, "Unicivilisation: The Dark Mountain Manifesto," 2009, http://dark-mountain.net/about/manifesto/.

42. Hamilton, *Requiem for a Species*, x–xi.

43. Ibid., 131–33, 211.

44. They are certainly not the first to articulate this lament. See especially McKibben, *End of Nature*; and McKibben, *Eaarth*.

45. On environmental grief, see especially Windle, "Ecology of Grief"; Pipher, *Green Boat*; and Eaton, "Navigating Anger."

46. His book ends with this paragraph: "Despair, Accept, Act. These are the three stages we must pass through. Despair is a natural human response to the new reality we face and to resist is to deny the truth. Although the duration and intensity of despair will vary among us, it is unhealthy and unhelpful to stop there. Emerging from despair means accepting the situation and resuming our equanimity; but if we go no further we risk becoming mired in passivity and fatalism. Only by acting, and acting ethically, can we redeem our humanity." Hamilton, *Requiem for a Species*, 226.

47. King, *Testament of Hope*, 225.

48. British Royal Society, *Geoengineering the Climate*.

49. See especially Robock, "Geoengineering May Be Bad." For a reflection on how Christian theology can inform debates about geoengineering, see Clingerman, "Between Babel and Pelagius."

50. Keith, *Case for Climate Engineering*.

51. Martin Lukacs, "World's Biggest Geoengineering Experiment 'Violates' UN Rules," October 15, 2012, www.theguardian.com/environment/2012/oct/15 /pacific-iron-fertilisation-geoengineering.

52. See especially Keith, *Case for Climate Engineering*; and Hulme, *Can Science Fix Climate Change?*

53. For a series of thoughtful essays on how geoengineering and decisions about it might be regulated, see Burns and Strauss, *Climate Change Geoengineering*.

54. See especially Lomborg, *Smart Solutions*.

55. Hamilton, *Earthmasters*.

56. King, *Where Do We Go from Here*, 172.

57. This perspective is developed in far more detail in chapter 10 of *Theological and Ethical Perspectives on Climate Engineering*, by Clingerman and O'Brien; and the rest of the chapters in that book offer additional important and challenging responses to the issue.

58. King, *Testament of Hope*, 252.

# 7

## Cesar Chavez and the Liberating Power of Sacrifice

> When we are really honest with ourselves, we must admit that our lives are all that really belong to us. So it is how we use our lives that determines what kind of men we are. It is my deepest belief that only by giving our lives do we find life. I am convinced that the truest act of courage, the strongest act of manliness, is to sacrifice ourselves for others in a totally nonviolent struggle for justice. To be a man is to suffer for others. God help us to be men!
> —Cesar Chavez, "Speech Ending 1968 Fast," in *Words of Cesar Chavez*

The civil rights movement and the struggle for racial and economic justice of course continued after the death of Martin Luther King Jr., as did non-violent protests. This is well reflected in the struggle for justice, health, and dignity among farmworkers, a movement that still cites Cesar Chavez as an iconic leader.

Chavez devoted his life to resisting the forces of structural violence oppressing farmworkers. As the world warms and the weather becomes less predictable, these workers' lives are likely to become more difficult. Growing seasons are changing, natural disasters threaten the livelihood of all who work on the land, and corporate growers are seeking more and more cost-saving measures. Thus, Chavez is still a vital witness for those who seek climate justice.

The quotation above is from a statement Chavez released in 1968 at the end of a twenty-five-day fast during his union's strike against grape growers in California.[1] He regularly incorporated fasting into his activism, and he encouraged those who worked with him to do the same, because he believed that such sacrifice helps to clarify purpose, to inspire discipline, and to foster commitment. He called those in his movement and those who supported it to take up a metaphorical cross, accepting sacrifices in order to make the world better. When made voluntarily, he argued, sacrifice is not a loss but a gain, and those who are willing to give their lives to the cause of justice will "find life."

International negotiations about climate change do not tend to explicitly mention sacrifice, but the topic is frequently implied in discussions of climate "debt" or "reparations." The idea is that the poor of the world, who have had relatively little impact on the climate, are owed something by the wealthy, who have disproportionately created the problem but do not tend to suffer its worst effects. This logic is straightforward: The developed world has less than 20 percent of the world's population but has produced more than 70 percent of anthropogenic greenhouse gases. Furthermore, the developing world has incurred more of the costs of climate change and is faced with higher risks and scarcer resources because of widespread poverty. Thus, the developed world owes a debt for contributing disproportionately to a problem whose costs fall more heavily on others.[2]

Such claims are often quickly rejected in one of two ways. The first dismisses the argument as counterproductive, because the best way to help those who struggle in a changing climate is for the rich to maintain a growing economy by expanding their own wealth. The second dismisses the argument as impolitic, because people in a consumeristic society will never be willing to listen to arguments or proposals that will cost them money. The first argument suggests that sacrifice is unnecessary, the second that it is impractical. In both cases, the assumption is that climate justice cannot be achieved by sacrifice. This chapter draws on the witness of Cesar Chavez to argue against such an assumption.

Admittedly, valorizing sacrifice is dangerous. The idea that sacrifice is restorative can be used by the privileged to ask the marginalized to give up more, and it can discourage or alienate those who seek moderate, pragmatic solutions to climate change. Though I take these dangers seriously, the following pages argue that privileged people are nevertheless called to make sacrifices. Those of us who have disproportionately changed the climate must give up some of the comforts, conveniences, and habits bought with greenhouse gas emissions. Chavez, who sought "a totally nonviolent struggle for justice" fueled by the voluntary sacrifice of his movement, can help twenty-first-century people articulate a morality of liberating sacrifice relevant to the changing climate.

## *LA CAUSA*

Cesar Chavez was born in 1927 on his family's farm near Yuma, Arizona. He lived there until age ten, when his family was evicted and moved to California to begin the migrant farmwork that defined the rest of his life. He once traced his willingness to stand up for workers' rights to this experience of settled life: "Some had been born into the migrant stream. But we had been on land, and I knew a different way of life. We were poor but we had liberty. The migrant is poor, and he has no freedom."[3]

After years of farmwork and a brief stint in the Navy, Chavez became a community organizer in 1952. He spent a decade registering Latino voters across the Southwest and then devoted himself to organizing farm laborers. In 1962 he and Dolores Huerta founded the National Farm Workers Association, which later became the United Farm Workers (UFW). At its first convention, Chavez was elected president and charged to advocate for a minimum wage law that would cover farmworkers. The motto of this convention, which has stuck with the movement ever since, was "*Viva la causa!*"

In 1965, Chavez led the UFW into its first extended strike, against grape growers in Delano, California. This strike evolved to include an extended public fast, an organized march across the state, and a national boycott. The campaign concluded with a more favorable contract between workers and grape growers, and Chavez spent the rest of his life seeking to maintain and expand the rights farmworkers had gained by raising national awareness about their working conditions.

Chavez was a Catholic of Mexican American descent, and his work grew out of his faith and his Church community. He told his biographer, "I don't think that I could base my will to struggle on cold economics or on some political doctrine. . . . For me the base must be faith."[4] Religion was also a tool for organizing. Preparing the UFW for its first action in 1962, Chavez had rose-flower workers place their hands on a crucifix and pledge to not break the strike.[5] At that event and all others to follow, the Virgin of Guadalupe was prominent as a symbol of hope and unity, of Catholic faith, and of Hispanic heritage. During his public fasts, many workers were stirred to religious devotion, holding Mass outside Chavez's lodging after some crawled from the highway on their knees in an act of penance and respect.[6] Although a few UFW leaders found these religious displays embarrassing, Chavez continued to publicly proclaim the Christian and Catholic foundations of his work throughout his life.[7]

But Chavez also sought to change religious institutions. He repeatedly called on the Catholic Church to use its power on behalf of the poor. He emphasized that it was not enough to provide charity and food baskets without also working to solve the social and political causes of poverty and oppression. In an essay titled "No More Cathedrals," he made the case this way: "Finally, in a nutshell, what do we want the Church to do? We don't ask for more cathedrals. We don't ask for bigger churches or fine gifts. We ask for its presence with us, beside us, as Christ among us. We ask the Church to *sacrifice with the people* for social change, for justice, and for love of brother. We don't ask for words. We ask for deeds. We don't ask for paternalism. We ask for servanthood" (emphasis in the original).[8]

Chavez believed that Christians are called to serve the poor, an idea consistent with the other four witnesses and the nonviolent tradition discussed in earlier chapters. For Chavez, this meant a particular commitment to those who work long days in farm fields. He frequently lamented that some workers were not even paid enough to afford the products they harvested: "They bring in so much food to feed you and me and the whole country and enough food to export to other places. The ironic thing and the tragic thing is that after they make this tremendous contribution, they don't have any money or any food left for themselves. And that's ridiculous.

Here they produce a tremendous amount of food and there is no food left for themselves and for their children."[9] Chavez's basic appeal was for recognition of human dignity, a demand that people should be fairly compensated for their work and that working people should have sufficient resources to feed their children.

Chavez insisted that farmworkers deserve better treatment by asserting the dignity of all people, and he used the same logic to insist that farmworkers must treat others with the same level of dignity. The only way to consistently assert the rights of workers as human beings, he argued, was for workers to respect the humanity of the growers against whom they struggled.[10] For the same reason, he discouraged attempts to overemphasize the singularly Hispanic identity of his movement. When many within the UFW began to stress that *la causa* was for *la raza*, he responded: "When you say *la raza* you are saying an anti-gringo thing, and our fear is that it won't stop there. Today it's anti-gringo, tomorrow it will be anti-Negro, and the day after it will be anti-Filipino, anti–Puerto Rican. And then it will be anti-poor-Mexican and anti-dark-skinned-Mexican."[11] So instead, he called for everyone to recognize and respect the humanity of all people, regardless of distinctions.

A commitment to human dignity was also demonstrated in *la causa* through a resolute commitment to nonviolence. When an airline pilot employed by growers sprayed a picket line with toxic pesticide in 1966, Chavez suppressed calls for revenge by quickly organizing a peaceful march from Delano to the state capitol in Sacramento. This outlet for anger successfully publicized the violence of growers but did not lead to further violence. Chavez insisted that the UFW's work could be successful and righteous only if it was nonviolent: "If to build our union required the deliberate taking of life, either the life of a grower or his child, or the life of a farmworker or his child, then I choose not to see the union built."[12] Having been profoundly inspired by the work of Mohandas Gandhi and Martin Luther King Jr., Chavez argued that the workers could best find their own dignity by not only respecting their opponents but also by learning to love them: "If we are full of hatred, we can't really do our work. Hatred saps all that strength and energy we need to plan."[13]

Chavez was first and foremost a union organizer who sought justice for farmworkers and led a people's movement for recognition and rights. However, with that work understood as primary, it is appropriate to note that Chavez was also an environmentalist. His work to honor human dignity included a commitment that everyone should have access to the natural resources necessary for survival and thriving. In a 1970 essay he lamented that the middle class of the United States was so excited about "a rocket to the Moon" but did not "give a damn about smog, oil leaks, the devastation of the environment with pesticides, hunger, disease."[14] In a 1976 speech he expressed his environmental impulse more positively, asking, "What could be more joyful than working to restore and preserve the sacredness of land, water, and air? For patriotism is not protecting the land of our fathers, but preserving the land for our children."[15]

The most explicit and extensive environmental aspect of Chavez's work came in his attention to the hazards of pesticides. The union's first grape boycott began in 1965, only three years after the publication of Rachel Carson's *Silent Spring*, and yet it already incorporated careful attention to the dangers of chemicals like DDT that threaten the health of farmworkers, food consumers, and the natural world.[16] Indeed, Chavez bragged that "the first ban on DDT, Aldrin, and Dieldrin in the United States was not by the Environmental Protection Agency in 1972, but in a United Farm Workers contract with a *grape grower* in 1967" (emphasis in the original).[17]

Building on this success, Chavez made pesticides the central subject of his final campaign, which was called "the Wrath of Grapes." In a 1989 speech he explained that "there is something more important to the farmworkers' union than winning better wages and working conditions. That is protecting farmworkers—and consumers—from systemic poisoning through the reckless use of agricultural toxics." He told the stories of grape workers' children who died of cancer at age five or were born without arms or legs, arguing that these were the direct result of their parents' exposure to pesticides. He called on his audience to boycott grapes and to make sacrifices for the sake of farmworkers and everyone else threatened by such pesticides. He also linked the issue of pesticides to broader environmental trends: "The problem is this mammoth agribusiness system. . . . The problem is the abandonment of cultural practices that have stood the test of centuries: crop rotation, diversification of crops. The problem is monoculture—growing acres and acres of the same crop; disrupting the natural order of things; letting insects feast on acres and acres of a harem of delight."[18]

In 1988 Chavez undertook a thirty-six-day "Fast for Life" to draw attention to this issue. The growers did not respond and refused to change their policies. However, Chavez garnered national attention and emphasized to audiences across the nation that he was doing penance and praying not only for the children of farmworkers but also for all who had been exposed to these poisons in their food. In his statement ending this fast, he called pesticide usage a "cycle of death that threatens our people and our world."[19]

Chavez died in Arizona in 1993, at the age of sixty-six years. Despite his eloquent writing, his national reputation, and his personal commitment to *la causa*, the UFW never met its initial goal of becoming a national union organizing all farmworkers. It began losing numbers and strength during Chavez's life, and a number of his critics blame his inflexible leadership style, his failure to build an organization that could thrive apart from his charisma, and his refusal to choose between creating a social movement and a labor union.[20] However, he has had a substantial influence on national culture. He remains a central Latino figure in US history, and generations of activists and politicians cite him, and in many cases their work with him, as a key inspiration.[21]

## THE LIBERATING POWER
## OF VOLUNTARY SACRIFICE

Chavez had a quiet voice and a relaxed speaking style, and he was never entirely comfortable in front of a crowd. His charisma came not from rhetorical force but from the strength of his personal witness. He created this image in part through his profound willingness to make sacrifices for the cause of justice.

The most iconic sacrifices were the three extended public fasts that Chavez undertook as a form of personal penance. Between twenty-one and thirty-six days in length, these fasts created a great deal of national attention, but he insisted that they were not for publicity. They were, instead, a personal necessity for his own spiritual health and dedication. This assertion is supported by the fact that Chavez conducted many other, private fasts throughout his life.[22] In part, he fasted because he found doing so to be liberating and clarifying and because he believed it helped him to experience life "on a different level," making him happier, wiser, and more patient.[23]

The sacrifice of fasting was a way for Chavez to limit the temptations of power and resist the comforts of contemporary life. He worked to convince all who worked with him to make sacrifices, as well, so that they would not fall into the trap of seeking easy answers and would instead build up the discipline to work for justice. In the course of his work to establish a fair wage for farmworkers, he also tried to lay the groundwork for a continued sense of cooperation and sacrifice, resisting the atmosphere of "superconsumerism" that shaped the lives of most wealthy people.[24] Even if farmworkers attained economic security through nonviolent protest and sacrifice, Chavez believed, they needed to maintain a strong sense of community and avoid entrapment in the lures of middle-class society.

Sacrifice was a central moral principle for Chavez because it resisted the problematic conveniences of mainstream culture and opposed the violent structures that told privileged people they could insulate themselves from the world's suffering. He insisted that true resistance to violence requires a willingness to take on some of this suffering: "There is no way on this Earth in which you can say yes to man's dignity and know that you're going to be spared some sacrifice."[25]

Chavez also believed that asking people to sacrifice was a way to ensure their commitment. At the UFW's founding, he insisted that monthly dues be required, despite the fact that workers were making very little and could not count on reliable daily work. He argued that if people were allowed not to pay because of economic hardship, then "they would never have a union because they couldn't afford to sacrifice a little more on top of their misery." Insisting on dues was also a strategic way to bring workers together: "Because, you see, they had an investment now. They were coming to the meetings, if for nothing else, but to see what was happening with their money that they were paying. And while they were there, we were taking advantage of educating them."[26] Sacrifice was a tool with which Chavez created community.

For the same reason, Chavez insisted that the UFW pay him and the other organizers only room, board, and $5 per week. Organizers who were willing to take this low wage worked out of dedication rather than self-interest, and Chavez believed this made them better organizers. This proved useful in the 1970s when the UFW competed with the Teamsters Union to represent grape workers. Chavez reported that his union won "with the numbers game" because he could hire ten organizers for what they paid one: "We don't have to worry about money. That's how things get done. . . . We'll organize workers in this movement as long as we're willing to sacrifice. The moment we stop sacrificing, we stop organizing."[27]

At the beginning of the Delano grape strike in 1965, Chavez and other leaders spelled out the key principles of their movement in the "Plan of Delano." Rather than discussing the specific tactics of strikes and boycotts, this document is a meditation on the suffering of farmworkers. It first frames their oppression as unjust and involuntary sacrifice imposed by growers, and then calls on the workers to sacrifice further for the sake of justice and a better future. Beginning from the premise that "our sweat and our blood have fallen on this land to make other men rich," the plan asserts:

> Our men, women, and children have suffered not only the basic brutality of stoop labor, and the most obvious injustices of the system; they have also suffered the desperation of knowing that that system caters to the greed of callous men and not to our needs. Now we will suffer for the purpose of ending the poverty, the misery, the injustice, with the hope that our children will not be exploited as we have been. They have imposed hungers on us, and now we hunger for justice. We draw our strength from the very despair in which we have been forced to live.[28]

The plan laments suffering, but then it calls for a sacrificial acceptance of more suffering. It asserts a willingness to give up meager pay and what little security came from grape picking in order to demand a more just arrangement with the growers.

Inherent in this foundational document is a distinction between involuntary and voluntary sacrifices. In much of the plan, Chavez and his coauthors critique the sacrifices imposed upon farmworkers, which kept them poor, hungry, and disenfranchised. Such involuntary sacrifices included the worst drudgeries of farmwork, which the UFW sought to prevent rather than celebrate. As Chavez told one biographer, "Every time I see lettuce, that's the first thing I think of, some human being had to thin it. And it's just like being nailed to a cross. You have to walk twisted, as you're stooped over, facing the row, and walking perpendicular to it."[29] This unnecessary sacrifice of farmworkers' health and well-being is a form of violence, which Chavez signaled by associating it with the crucifixion of Christ. Such involuntary sacrifice was the problem the UFW worked to solve.

However, Chavez taught that liberation from such violence can be found in voluntary sacrifice. Thus, the Plan of Delano also called on workers to choose redemptive and restorative sacrifices, a choice that its authors argued would contribute to self-worth, dignity, and a better future. When workers gave their resources to the movement, they were more invested in its work; when organizers sacrificed lucrative careers, they became more flexible and less beholden to the wealthy; when Chavez went without food, he became clearer and more attentive. Such sacrifices reflected both an organizing strategy and a spiritual practice.[30]

## THE IMPORTANCE AND LIMITS OF SACRIFICE

Those concerned about climate justice should learn to think and communicate about sacrifice, and Chavez is a helpful resource. We should take seriously the involuntary sacrifices already being made by the peoples of Bangladesh, Sudan, Kivalina, and those in many other parts of the world. We must admit that the best way for privileged people to help liberate such people from involuntary sacrifices is by making our own voluntary sacrifices. However, we also need to acknowledge that the world's injustices and structural violence make sacrifice a morally fraught category.

### Sacrifice Is Necessary and Possible

Sacrifice is most often raised in public discourse about climate change in order to argue against it. For example, in 2014, the managing director of the World Bank, Sri Indrawati, argued that the global community must take action on climate change but that this action "does not have to come at the expense of economic growth." The headline of her op-ed was: "Climate Action Does Not Require Economic Sacrifice."[31] This is a reassuring perspective for many, and she backed it up with careful economic data. However, the previous chapters of this book have told a different story. The witness of Jane Addams makes clear that economic and political change must accompany any attempt to end structural violence. The witness of

John Woolman suggests that such change must begin with privileged people altering their lifestyles, sacrificing some of their comforts. Responding morally to climate change means being willing to change the structures in which privileged people and other human beings live. This will feel like a sacrifice to those who live comfortably in the existing system.

The social ethicist Anna Peterson offers a powerful counterargument to the common insistence that sacrifice is unnecessary. As a scholar of religion, she notes that rituals of sacrifice are ubiquitous across every human society, and this raises suspicion about any claim that no one needs to sacrifice. Any transition toward a sustainable society will require sacrifices from affluent Westerners, Peterson argues, and she further emphasizes that environmentalists should identify and explore "the sacrificial logic of everyday life in American culture," asking people "what we are willing to give up, and why." She points to the concrete sacrifices that contemporary US society expects from its citizens:

> We take for granted the fact that we "have to" sacrifice excellent health care or public education, or clean air and water, or safe highways, or a number of other public (and private) goods, not because we do not value them, and not even because we value the goods for which they are sacrificed (usually lower taxes or corporate profits). We take these sacrifices for granted because they are not called sacrifices; they are simply part of "the way things are."[32]

Awareness of structural violence calls for a close examination of anything that the existing power structure frames as inevitable and unchanging. This means carefully considering the possibility that some economic growth should be sacrificed in order to respond to the violence of climate change. When a leader of the World Bank insists that economic growth is nonnegotiable, it is important to ask hard questions about what sacrifices are already being made to ensure that such growth continues.

To frame sacrifice as unnecessary is to ignore the fact that sacrifices are already being made. In their book *The Environmental Politics of Sacrifice*, the political scientists Michael Maniates and John Meyer call attention to the "hidden and coerced sacrifice already borne by those most vulnerable to climate disruption."[33] As waters rise in Bangladesh and Kivalina, people's homes are being sacrificed. Some people are losing their lives. Such involuntary sacrifices call for a liberation that can only come through voluntary sacrifice by those who can afford it.

Another argument against sacrifice suggests that even if it is a good idea, mentioning it is counterproductive because people will never voluntarily give up privilege. Consider the observation of a *USA Today* editor on Earth Day in 2014: "The vast majority of us won't sacrifice our high-consumption lifestyles, put jobs at risk or pay a high price to protect the environment. . . . We'll continue to mark Earth Day, then drive home, turn on the TV, take a swig of water from our plastic bottles and check out the latest post on our Facebook pages."[34] Whether sacrifice is necessary or not, this perspective asserts, it will never happen.

Again, the witnesses considered in this book offer a counterargument. Martin Luther King Jr. chose to put himself at risk in order to live out his commitment to justice. Dorothy Day gave up economic security and made a commitment to live among the poor. These sacrifices were made by rare witnesses, but they were supported by thousands of others who made smaller sacrifices of wealth, time, and attention. Privileged people can make sacrifices. We know this is true because it has happened.

Maniates and Meyer criticize mainstream environmental groups in the industrialized world that "treat sacrifice as an idea to be avoided at any cost, lest environmentalism become identified as a movement of deprivation and hair shirts."[35] To dismiss sacrifice because it will not sell is to ignore the realities of suffering and sacrifice all over the world among the poor and marginalized. Anyone who seeks climate justice, who hopes to respond morally to the structural violence of climate change, must take the involuntary sacrifices of others seriously and consider the possibility that it is time to make voluntary sacrifices in response.

## Sacrifice Is Dangerous

Still, calls for "sacrifice" must seriously consider this term's limitations and dangers. One perspective on these limitations and dangers is demonstrated in the quotation from Cesar Chavez that began this chapter: "To be a man is to suffer for others. God help us to be men."[36] In 1972 Chavez likely considered this to be an inclusive statement, meaning to incorporate all people with the word "man." However, the implied identification of sacrifice with masculinity resonates with feminist critiques that have questioned whether, in a patriarchal society, sacrifice is as noble for women as it is for men.

In an iconic essay, the theologian Valerie Saiving observes that male Christian thinkers frequently "identify sin with self-assertion and love with selflessness." This view valorizes sacrifice, associating Christian love with self-abnegation and sin with the prideful inflation of the self. Saiving then argues that though this is an accurate account of many mens' moral lives, it does not reflect the experience of women. Because women have been socialized by patriarchal cultures to care for others rather than themselves, they face a different temptation, to "give too much of herself, so that nothing remains of her own uniqueness; she can become merely an emptiness, almost a zero, without value to herself, to her fellow men, or, perhaps, even to God."[37] It is dangerous to instruct such women that sacrifice is virtuous because it can end up telling those who have been marginalized that they should not stand up against the structures that marginalize them. Saiving suggests that men in a patriarchal society should indeed learn to limit themselves and surrender some desires but that such an ethics based exclusively on male experience must be questioned and balanced by an ethics based upon womens' experiences. This ethics would recognize the importance of self-assertion and self-defense for women rather than or in addition to sacrifice.

Barbara Hilkert Andolsen makes a similar point by arguing against the uncritical celebration of sacrificial love in Christian ethics: "The virtues which theologians should be urging upon women as women are autonomy and self-realization. What many male theologians are offering instead is a one-sided call to self-sacrifice which may ironically enforce women's sins."[38] The same argument must be made for people of color, the poor, and other marginalized groups, whom the privileged have no right to ask to sacrifice on behalf of a system that already denies them agency and power.

These critics rightly reject involuntary sacrifice—sacrifices forced upon those without agency. This is a vital caution for anyone who advocates sacrifice in any form. However, it does not forgo the possibility of a different kind of sacrifice that liberates people rather than constrains them. For example, the ecofeminist Mary Grey takes the critiques of sacrifice seriously, but she then goes on to argue that sacrifice and austerity are essential as "the only means of effective resistance to an unjust world order." She urges women and men who benefit from a world of excessive consumption to choose a different path by embracing some sacrifices. Giving up some of the privileges that come from economic dominance and industrialism, she argues, is the only way to affirm "life for all, joy and justice for all, sustainable living for all."[39]

Sacrifice is dangerous, so a moral response to climate change cannot simplistically call for universal sacrifice. Suffering is not always liberatory. Those who struggle to survive and assert themselves should not be called upon to suffer more, nor to give away what little they have. However, sacrifice is a vital practice for privileged citizens of the industrialized world. Privileged people concerned about climate justice must, therefore, find ways to sacrificially turn away from those systems and structures that have served our desires at the expense of other people, of ecosystems, and of future generations.

## CLIMATE CHANGE AND SACRIFICE

Cesar Chavez called on his audiences to give up the destructive aspects of contemporary life, to sacrifice comforts and even standards of success if they were based upon the suffering of others. In other words, he called on his followers to resist structural violence by withdrawing from violent structures. Following this calling may feel like a sacrifice if it means skipping a vacation, forgoing a favorite food, leaving a job, or accepting a lower income. But Chavez promised that the freedom and energy that come from sacrifice can make it liberatory. This is a crucial idea that can inspire the movement for climate justice, which should call its privileged members to sacrifice destructive ways of life for the sake of those whose lives and well-being are already being involuntarily sacrificed.

### Resisting Involuntary Sacrifice

The first lesson to be drawn from Chavez is that it is immoral to force involuntary sacrifices on the poor and marginalized. He protested against the meager wages paid to farmworkers and the indignities of unrelieved backbreaking labor. Climate justice advocates should continue this protest today. Indeed, this protest may be even more urgent than in Chavez's time, given that UFW veteran Marshall Ganz reports that "in real dollars, the nearly 400,000 farmworkers employed today by California's $30 billion agricultural industry earn wages 20–25 percent below those paid in the late 1970s."[40] Pesticide use also remains a prevalent and understudied problem. Chavez's witness calls for resistance against such impoverishment.

Climate change makes life even more difficult for farmworkers. Heat stroke has long been a serious threat in the fields, and increasing extreme weather can increase that danger. Changes in growing conditions may well lead to the use of more chemical fertilizers and pesticides, and those who apply these chemicals and work closest to the crops face the most dangers.[41] Poor farmworkers are, sadly, another example of the unfair burdens of climate change placed upon those who have contributed least to the problem.

Climate change will demand sacrifices in the twenty-first century, perhaps from every human being on Earth. However, such sacrifices should not come first or foremost from those already struggling to survive. Chavez's witness calls for a resolute stand against a system that "caters to the greed of callous men" rather than to the needs of those who struggle to feed and shelter their families. Involuntary sacrifice is not redemptive or constructive. It is, instead, a form of violence.

In this sense, Chavez's witness is consistent with the feminist ethicists cited above on the dangers of sacrifice. Like them, Chavez insists that the privileged have no right to ask for sacrifice from those with less.[42] Instead, he asked for sacrifice first from people with whom he was in community and those with more power than himself. Identifying himself as a farmworker and organizer, he asked others in those communities to dig deep and make sacrifices, paying union dues and taking reduced wages. He ensured that his own lifestyle matched that of the workers and so remained on an even footing with them. He asked those with more wealth than himself to make sacrifices, exemplified in his call for the Church to sacrifice grandiosity for the sake of social justice. He also traveled widely to ask privileged people across the country to boycott grapes and to fast in solidarity with suffering farmworkers. He asked a college audience in 1989 to "join the many hundreds who have taken up where my fast ended—by sharing the suffering of the farmworkers—by going without food for a day or two days or three."[43]

## The Call to Consume Less

For Chavez, the danger of involuntary sacrifice is matched by the power of voluntary sacrifice. This, too, is a vital lesson for privileged people today. We should ask our own communities and those with more power than ourselves to make voluntary sacrifices. Citizens with economic security in the industrialized world should consider giving up comforts in order to reduce our carbon footprints. We should undertake energy or dietary fasts that reduce dependence on destructive fossil fuel and agribusiness infrastructures. Wealthy consumers should follow the example of the UFW grape boycott and forgo products with particularly destructive environmental records. Learning from Chavez, we may discover that such sacrifices can help us to feel less imprisoned by consumer culture, free to live up to our truest desires and ideals. As discussed in chapter 3, privileged people concerned about climate change should eat lower on the food chain, make careful choices about our transportation, and consume energy as responsibly as possible. Chavez complements the witness of John Woolman, agreeing that such sacrifices can in fact bring more joy and satisfaction to people's lives.

This echoes the call for voluntary poverty made on the basis of Dorothy Day's witness, as described in chapter 5. The ecofeminist theologian Sallie McFague, who picks up this call from Day, writes that wealthy people in the industrialized world "cannot love our neighbors—neither the human ones nor the earth ones—unless we drastically cut back on our consumption."[44] So, McFague suggests that privileged people in the twenty-first century must learn the gift of "self-emptying," of "self-restraint, giving space to others, pulling back, saying 'enough.' "[45] To sacrifice, she explains, is to put the community above oneself, to recognize that one's neighbors who suffer deserve the basic necessities of life from those who have more than they need. Like Chavez, she insists that such sacrifice is ultimately joyful, turning away from the never-ending and unsatisfying cycle of overconsumption and instead embracing "a vision of sustainable and just planetary living."[46]

Chavez emptied himself—literally and figuratively—by fasting for his movement, and in so doing he found a peace and clarity that inspired him to continue devoting his life to justice. However, his example also extends beyond the personal. He worked hard to make political changes, not only calling on individuals to make voluntary sacrifices but also seeking to restructure economic and political systems so that the entire nation would learn to respect the dignity of farmworkers. In a democracy, widespread public sacrifice can be considered voluntary when a majority decides it is a good idea. Thus, the movement for climate justice should use politics to help privileged people make sacrifices. Activists should seek laws that require everyone to pay more for the privilege of burning fossil fuels, so that they will be burned less often. Activists should demand regulations to increase the price of industrially produced meat and other foods that disproportionately contribute to climate change and thus encourage the production of more locally sourced and lower-impact foods. The movement should also work to see that the money society saves from these sacrifices does not go into the investment portfolios of the rich but rather into projects that benefit those who already suffer the most from a changing climate. If political systems are engaged carefully, such sacrifices can change the lifestyles of the privileged without becoming undue burdens on those who already struggle.

## The Energizing Power of Sacrifice

In a world dominated by the logic of the market, such calls for giving up wealth and comfort will not be popular. Thus, the central idea quoted from Chavez at the beginning of this chapter is crucial: The only way to be fully human is to say yes to sacrifice out of respect for the dignity of others. Chavez promises that in making sacrifices, people will discover true and deep values, rewards that are far more worthy than money and consumer goods.

Chavez lived out this belief by insisting that even poor workers must pay dues and union organizers should be paid very little. Other unions treated membership and organizing as purely economic transactions—members expected to get their money's worth, and employees expected to be paid what the market dictates. But Chavez worked to build a union based on community and a commitment to human dignity. The value of the union was the sense of belonging and inclusion it provided, and organizers were expected to commit to the cause rather than their paycheck.

Privileged people seeking climate justice need not copy every concrete strategy that Chavez used with UFW members and workers. It is not our place to tell the poor how to spend their resources, and those who can afford it should support a living wage for all who do the work of activism and organizing. However, the broader point is worthy of careful study: People are more committed and more energized when they have sacrificed.

The communications scholar Shane Gunster makes a related point critiquing the rollout of British Columbia's carbon tax. This policy, which was instituted in 2008, added additional taxes to any fossil fuels burned for home heating, transportation, or electricity and is frequently cited as a model for political attempts to slow climate change. The tax was designed to be revenue neutral, such that every cent raised is channeled back to the public in the form of lower income taxes. Gunster characterizes this decision as based on the belief "that a fickle, narcissistic public can only be brought on side in the fight against global warming by appealing to their personal self-interest."[47] He critiques this assumption, saying that political leaders ignored citizens' hunger for ways to take genuine action on climate change.

By emphasizing that British Columbians need to make no sacrifice for a carbon tax, the government communicated that citizens need not do much about climate change. Gunster suggests that this was a mistake. It ignored "how the process of *collectively* mobilizing the immense social, economic, and political resources of society to tackle environmental issues after decades of neglect and indifference is enormously appealing to large segments of the population." People understand that climate change is a huge and serious problem, and many would feel empowered by the chance to do something significant in response. In this context, perhaps asking for widespread sacrifice is a way to meet a genuine need. Perhaps giving something up for the sake of the climate will be not only "intellectually compelling, but also emotionally attractive, carrying in its wake the genuine pleasure of working and acting together to make a better world."[48]

Politicians in British Columbia did not ask for a genuine sacrifice, so it is impossible to know how citizens might have reacted. But Gunster's suggestion that people are capable of such goodness when they are asked is a hopeful argument, and it is consistent with Chavez's insight about the importance of sacrifice in solidifying commitment. The movement for climate justice should be calling for sacrifice from those who can afford it because people are likely far more open to such requests than conventional wisdom assumes.

## CLIMATE DEBT AND REPARATIONS AS
## LIBERATING SACRIFICE

A concrete call for sacrifice from privileged people comes in the demands for climate reparations to repay a climate debt. One epicenter of such calls is the nation of Bolivia, which sadly provides further examples of the structural violence inherent in climate change. Bolivia is home to 20 percent of the world's tropical glaciers, which are rapidly shrinking. This creates both floods and droughts—floods when excess water comes from mountaintops far more rapidly than it used to and droughts when the reduced ice pack provides inadequate water in dry months.

A 2009 Oxfam report tracks the many ways Bolivians have already been affected by climate change, with more extreme weather events, shrinking water supplies, and endangered food sources. Lucia Quispe, a Bolivian farmer in the rural community of Khapi, told the Oxfam researchers that "it does not rain when it should any more. At any moment, there might be clouds and the rain falls. Before, there was a season for rain, and a season for frost, and a period of winter. [Now,] we get more illnesses, more colds, more coughs, because of the sudden changes in the weather." When asked how she felt about her children's future, Quispe responded, "Sad, and very worried. If we don't have water, what are we going to live from? There's no life without water."[49]

As one of the poorest nations in Latin America, Bolivia can ill afford such disruptions. Adaptation to climate change is economically challenging or impossible for many Bolivians. And, in a trend that is disturbingly familiar, the nation has contributed relatively little to the problem of climate change from which it suffers. In 2012 Bolivians were responsible for 0.3 percent of the world's greenhouse gas emissions. By comparison, the United States was responsible for 13 percent.[50]

Bolivia has the highest percentage of indigenous peoples of any nation in South America, and these communities are also profoundly threatened. Consider the Uru Chipaya, a community on the Lauca River that has maintained a consistent way of life for four thousand years, outlasting the Incan Empire and sustaining its culture through Spanish colonization. But this culture now faces extinction because the river by which it has always defined itself is drying up. The Uru Chipaya insist that while individuals might survive and migrate elsewhere, its culture cannot endure without the Lauca River.

In 2009 the Bolivian ambassador to the World Trade Organization, Angélica Navarro Llanos, stated that the people of Bolivia are owed a climate debt.[51] The nation's economy, the traditional cultures of its indigenous peoples, and the very lives of its citizens are threatened by a problem that was created and continues to be disproportionately caused by others. Navarro Llanos argued that her nation deserves reparations for the damage climate change is causing.

This proposal was controversial. Responding to Navarro Llanos and other nations making similar requests, the United States' special envoy for climate change refused to accept responsibility for the harm of climate change. He argued that because US citizens and officials had not realized the harm they were causing, they do not owe anyone anything: "I actually completely reject the notion of a debt or reparations or anything of the like. . . . For most of the 200 years since the Industrial Revolution, people were blissfully ignorant of the fact that emissions caused a greenhouse effect."[52] Critics questioned the claim that the United States was "blissfully ignorant" all the way up until 2009. Even if such ignorance were defensible, it would raise the question of whether the United States owes some reparation for the harms its emissions have caused in the years since 2009.

Another rejection of climate debt comes from the political philosophers Jonathan Pickering and Christian Barry. Unlike Stern, they concede that there is a moral logic to the call for reparations, but they then suggest that this is not a strategic case to make in the contemporary global climate. They worry that demanding a debt be paid or harm be repaired will inspire defensiveness in the industrialized world, whose citizens will seek to resist claims of guilt rather than genuinely engage the shared global problem of climate change.[53] Again, there are two arguments being made against a call for sacrifice: that it is unfair and that it is impractical.

The witness of Cesar Chavez suggests otherwise—that sacrifice can be a moral response to violence that is empowering rather than divisive, if it is approached thoughtfully. This witness calls concerned people to take meaningful action in response to the involuntary suffering of the poor and oppressed in Bolivia. Chavez's *causa* demonstrates that sacrifice creates community, and that people are more likely to believe in things for which they have made sacrifices. By this logic, a genuine discussion of climate debt might invite citizens of the industrialized world to engage with the suffering of others in a genuine way rather than ignoring it. Indeed, perhaps the reason that privileged people talk so little about the struggles of Bolivia in a changing climate is that they feel powerless to do anything about it.

A meaningful sacrifice that sent some of our wealth to those who are suffering most could allow privileged people to honestly face the structural violence of climate change. Along these lines, the legal scholar Maxine Burkett advocates reparations not only as a way to compensate those who have suffered from climate change but also as a way to "engage the globe—particularly those in the developed world—in the great ethical challenge posed by climate change." The idea of paying for harm would allow privileged people to publicly acknowledge what most of us already know but rarely talk about: Our lifestyles contribute to structural violence. According to Burkett, the power of reparation language is that it emphasizes the moral dimensions of climate change. It moves the discussion away from purely scientific and economic issues and into the realm of morality. As she explains, "A discussion distinct from 'caps,' 'trades,' and 'costs to the average consumer' will help to illuminate suffering of the climate vulnerable, and the developed world's understanding of its own responsibility."[54]

National and international debates about climate debt and climate reparations will go on for many years to come. They may never be satisfyingly resolved. But concerned people committed to resisting structural violence should speak up for nations like Bolivia. Bolivians are owed a debt, and we should encourage our fellow citizens to consider how paying such debts could be a liberating sacrifice rather than a grueling punishment. Indeed, a discussion of climate reparations might be the only way to have an honest and mature conversation about climate change.[55]

Following the witness of Chavez, concerned people should embrace the chance to be fully human by recognizing the suffering of others, taking our fair share of responsibility for this suffering, and making sacrifices to alleviate it. Paying the debts that we and our ancestors have incurred will be a sacrifice, but it will contribute to the liberation of those who suffer from the violence of climate change.

## NOTES

Most who write about Cesar Chavez use accents over a letter in each of his names—
"César Chávez"—which would be correct Spanish. I do not write it this way because
he himself wrote "Cesar Chavez." I learned this from Luis León, who notes that
Chavez has a "right to misspell his name consistently in a lifetime of signing it."
León, *Political Spirituality of Chavez*, 35. To avoid distracting the reader, this style is
also followed for the works cited.

For biographies and analyses of Chavez, see especially Levy and Chavez, *Cesar
Chavez*; Del Castillo and Garcia, *Cesar Chavez*; Ferriss, Sandoval, and Hembree,
*Fight in the Fields*; Etulain, *Cesar Chavez*; Dalton, *Moral Vision of Cesar Chavez*;
Shaw, *Beyond the Fields*; and Pawel, *Crusades of Cesar Chavez*.

1. Chavez, *Words of Cesar Chavez*, 167.
2. Third World Network, "Climate Debt: A Primer," June 2009, www.twn.my/title2
   /climate/briefings/Bonn03/TWN.BPjune2009.bonn.02.doc. For an articulation
   of similar arguments that focuses more on domestic environmental reparations
   in the United States, see Collin and Collin, "Environmental Reparations"; and
   Harvey, "Dangerous 'Goods.'"
3. Chavez, *Words of Cesar Chavez*, xviii.
4. Levy and Chavez, *Cesar Chavez*, 27.
5. Watt, *Farm Workers and Churches*, 71.
6. Moffett, "Holy Activist, Secular Saint," 106.
7. Prouty, *Struggle for Social Justice*, 24. Luis León makes an interesting argument
   that Catholicism should be understood as just one aspect of Chavez's religious
   life and that *la causa* itself is best understood as its own "quasi-religion of resistance." León, *Political Spirituality of Chavez*, 157. I agree with León that Chavez
   chose to ignore some Catholic doctrine and brought other religious rituals and
   ideas into conversation with his Catholic background. However, this seems to
   me to be entirely consistent with the faith of most Catholics, so I continue to
   refer to him primarily as a Catholic thinker and activist.
8. Chavez, *An Organizer's Tale*, 87.
9. Chavez, *Words of Cesar Chavez*, 86–87.
10. Matthiessen, *Sal Si Puedes*, 115.
11. Ibid., 143.
12. Chavez, *Words of Cesar Chavez*, 36.
13. Levy and Chavez, *Cesar Chavez*, 196.
14. Chavez, *An Organizer's Tale*, 83.
15. Chavez, *Words of Cesar Chavez*, 92.
16. Carson, *Silent Spring*.

17. Chavez, *Words of Cesar Chavez*, 144.

18. Ibid., 144, 147.

19. Ibid., 168–69.

20. For discussions of the UFW's decline, see Shaw, *Beyond the Fields*, chap. 10; Ganz, *Why David Sometimes Wins*, epilogue; and Pawel, *Union of Their Dreams*, parts V and VI.

21. Along these lines, the activist Randy Shaw argues that the key legacy of the UFW comes from the training it gave its organizers, many of whom went on to work in other organizations and other movements. Shaw, *Beyond the Fields*, 5.

22. One report suggests that he fasted almost one in five days for much of his adult life. Hribar, "Social Fasts of Cesar Chavez," 332.

23. Levy and Chavez, *Cesar Chavez*, 350.

24. Chavez, *Words of Cesar Chavez*, 144.

25. Levy and Chavez, *Cesar Chavez*, 539.

26. Chavez, *Words of Cesar Chavez*, 66.

27. Ibid., 69–70.

28. Ibid., 16–17. For a complementary analysis of this text emphasizing its focus on human dignity as well as sacrifice, see Dalton, *Moral Vision of Cesar Chavez*, 87–89.

29. Levy and Chavez, *Cesar Chavez*, 74.

30. This combination of a broad strategy with a personal discipline is typical of Chavez, as the ethicist Frederick John Dalton insightfully observes. Dalton characterizes Chavez's morality as a creative synthesis of "liberation ethics" and "character ethics," thus bridging concerns for social justice and personal virtue. Dalton, *Moral Vision of Cesar Chavez*.

31. Sri Mulyani Indrawati, "Climate Action Does Not Require Economic Sacrifice," 2014, http://blogs.worldbank.org/voices/climate-action-does-not-require-economic-sacrifice.

32. Peterson, "Ordinary and Extraordinary Sacrifices," 96, 105.

33. Maniates and Meyer, *Environmental Politics of Sacrifice*, "Conclusion: Sacrifice and a New Environmental Politics," 315.

34. Owen Ullmann, "Voices: On Earth Day 2014, a Climate Change Challenge," *USA Today*, April 21, 2014, www.usatoday.com/story/news/nation/2014/04/21/earth-day-environment/7905743/.

35. Maniates and Meyer, *Environmental Politics of Sacrifice*, "Conclusion," 313.

36. At Chavez's funeral in 1993, the end of this iconic quote was changed so that it read, "To be human is to suffer for others. God help me to be human"; Rodriguez, *Darling*, 138. Committed as he was to the dignity of all human beings, Chavez would probably not have objected to a change that made his statement more inclusive. However, the fact that it was originally written with exclusive

language signals the possibility that there is a gendered character to such a call for sacrifice, that suffering for others in order "to be a man" is not universally advisable.

37. Saiving, "Human Situation," 26, 43.

38. Andolsen, "Agape in Feminist Ethics," 151.

39. Grey, *Sacred Longings*, 188. See also Mercedes, *Power for*.

40. Ganz, *Why David Sometimes Wins*, 239. See also Oxfam America, *Like Machines*.

41. See, e.g., Northwest Public Radio, "How Farmworkers Experience a Warming Climate," September 27, 2013, http://earthfix.opb.org/communities/article/how-migrant-workers-experience-a-warming-climate/.

42. This is not to say that Chavez should be seen as an uncomplicated feminist hero. He did not fully account for the burdens his sacrifices placed on his wife and his family, and he was not always welcoming to women's perspectives, even among the leadership of the UFW. See especially Pawel, *Crusades of Cesar Chavez*.

43. Chavez, *Words of Cesar Chavez*, 149.

44. McFague, *Life Abundant*, 22–23.

45. McFague, *Blessed Are the Consumers*, 17.

46. Ibid., 74.

47. Gunster, "Self-Interest, Sacrifice, and Climate Change," 196.

48. Ibid., 202, 210.

49. "Bolivia: Climate Change, Poverty, and Adaptation," 2009, p. 39, www.oxfam.org/sites/www.oxfam.org/files/file_attachments/bolivia-climate-change-adaptation-0911_4.pdf.

50. "CAIT Climate Data Explorer," http://cait.wri.org/historical/Country%20GHG%20Emissions?indicator[]=Total%20GHG%20Emissions%20Excluding%20Land-Use%20Change%20and%20Forestry&indicator=Total%20GHG%20Emissions%20Including%20Land-Use%20Change%20and%20Forestry&year=2012&sortIdx=NaN&chartType=geo. These numbers include the greenhouse gas equivalents of land use and forestry. If these figures are taken out and only gas emissions are considered, Bolivia's share drops to 0.1 percent, while the US share rises to 14 percent.

51. Llanos, "Climate Debt."

52. Andrew C. Revkin and Tom Zeller Jr., "US Negotiator Dismisses Reparations for Climate," *New York Times*, December 9, 2009.

53. Pickering and Barry, "Concept of Climate Debt."

54. Maxine Burkett, "Climate Reparations," 511.

55. Ta-Nehisi Coates eloquently makes this argument with regard to reparations for slavery and racism in the United States: "I believe that wrestling publicly with these questions matters as much as—if not more than—the specific answers that might be produced. An America that asks what it owes its most vulnerable

citizens is improved and humane. An America that looks away is ignoring not just the sins of the past but the sins of the present and the certain sins of the future. More important than any single check cut to any African American, the payment of reparations would represent America's maturation out of the childhood myth of its innocence into a wisdom worthy of its founders." Ta-Nehisi Coates, "The Case for Reparations," *The Atlantic*, May 21, 2014, www.theatlantic.com/features/archive/2014/05/the-case-for-reparations/361631/.

# Conclusion

## What Can We Do?

Climate change is an impossibly hard problem. To fully understand it would require many distinct ways of knowing, almost certainly more than one person could ever master. Continued changes and unpredictability are guaranteed by the chemistry of the atmosphere and the inertia of industrial human civilization. The effects of these changes are unjust, suffered by those with the least power to prevent them—poor and marginalized human beings and nonhuman creatures.

So, what can we do? This is a dangerous question. Too often, we ask what to do in the face of a complex problem but are only open to easy solutions. We look for a way to feel as if we have done something—donated money, bought a more efficient appliance, or wagged a finger at someone doing more harm than ourselves—in hopes that it will be enough. Or we look for the promise of an easy answer through technology or politics, hoping we can feel righteous or optimistic while trusting others to solve the problem. If "what can we do" means seeking an easy solution to climate change, there is no answer. Climate change is a wicked problem; we will not solve it.

But of course that does not mean there is nothing to do. To the contrary, this book has argued that there is a great deal that privileged people in the industrialized world can and should do if we learn from five witnesses who demonstrate resistance against structural violence.

There is danger in valorizing such witnesses. It is easy to think that these impressive women and men are different from you and me, that they are a special class of person with whom we have little or nothing in common. The theologian Brian Mahan captures this well, writing about the danger of "round[ing] up 'the usual suspects'—Gandhi, Mother Teresa, and Martin Luther King Jr."—to teach about living a meaningful life. It is easy to respond dismissively, saying, "Yes, they lived lives like that. And we really all love them. They even make us cry sometimes. But after all, there's only three of them."[1] We could add a few more names to Mahan's list but must recognize the impulse to which he points—celebrating the wisdom and action of amazing people can make the rest of us feel distant and ineffectual rather than empowered. The witnesses cited in this book are impressive, but after all there are only five of them. Surely the rest of us cannot be expected to do what they did?

No, we cannot. Of course not. We live in a different time, in a different world. We should not seek to be Chavez or King or Day or Addams or Woolman. We should be ourselves, in our time. But this does not let us off the hook. We cannot be them, but we can be like them insofar as that means devoting our lives to resisting violence without resorting to violence; I have argued throughout this book that we should do so. In order to make this more possible, I now seek to humanize these five witnesses slightly, to show the ways in which they are enough like us, and different enough from one another, that it is reasonable to use them as guides as we seek new paths forward. These witnesses teach us that resistance can be part of a fulfilled and joyful life, that there are diverse paths of resistance, and that resistance is never a finished project.

## JOYFUL LIVING

Having spent years with the words and stories of these five witnesses, there is one characteristic I fear I have not captured well in the preceding chapters. I have worked to represent the ways they committed themselves to nonviolent resistance against the dehumanizing and destructive forces around them and how they creatively and nobly sacrificed and struggled to make the world better. But I worry that the preceding chapters do not convey the fact that each one also took genuine pleasure in that work, enjoying life and the world even as they resisted violence.

Cesar Chavez named his German shepherds "Boycott" and "Huelga" (the Spanish word for "strike"), which reflects his down-to-earth sense of humor about his work. He frequently made gentle jokes at his own expense or against his political opponents. During the United Farm Workers' first strike, he humbly told his biographer that he and the members of his union were still learning to nonviolently love grape growers and noted that "I think we've learned how not to hate them, and maybe love comes in stages." He then wryly added, "Of course, we can learn how to love the growers more easily after they sign contracts."[2] In public speeches during that strike, while telling audiences to boycott grapes and wine produced by the Gallo corporation, he joked about a way to find out if wine was made by that company: "Drink the stuff, and if you get sick, that's also Gallo wine."[3]

Dorothy Day had a wickedly ironic sense of humor that she used to disarm overly serious and self-important people. When a deeply concerned wealthy woman asked Day what could be done about the rampant poverty in New York City, Day dryly replied, "Let's blow it all up!"[4] At the age of eighty-one years, she reported on a meal at the Catholic Worker House in her column, writing of "hard, baked potatoes for supper, and cabbage overspiced," and adding this commentary: "I'm in favor of becoming a vegetarian only if the vegetables are cooked right."[5]

Martin Luther King Jr. also loved good food, and he frequently enjoyed the company of family and friends around a table piled with southern delicacies. The last hour of his life was spent preparing for such a meal with close friends. A Memphis pastor's wife had gathered the best cooks from her church to prepare roast beef, potatoes, macaroni and cheese, greens, sweet potatoes, pies, and ice cream. One of King's final conversations was with the hotel owner who had served him lunch, and he promised her that if his dinner was not good "like that catfish we had, I'm going to come back and eat here." Soon afterward, a friend was teasing that King was putting on weight and his clothes did not fit, and King laughingly agreed.[6] The horrible tragedy of King's assassination came next, but his last moments were filled with camaraderie and joyful anticipation of a good meal. This is a sign of the life he lived, appreciating the world even as he sought to make it more just.[7]

John Woolman has left a more serious record of his thought and life, and I do not know any stories of jokes or laughter from him. But his seriousness of purpose came from a desire to live happily. He believed that the most joyful people are those who truly and humbly listen to the voice of light and wisdom inside themselves. In his journal, he wrote, "Some glances of real beauty may be seen in their faces who dwell in true meekness. There is . . . right order in their temper and conduct whose passions are fully regulated."[8] In other words, Woolman sought to avoid greed, frivolity, and injustice because he believed they would ultimately be unsatisfying. His quest for a humble, peaceful life was a quest to find joyful balance.

Jane Addams also had a serious public presence, but in private she was a cheerful woman who enjoyed life. This is clearest in her deep friendships, which shaped and sustained her work while bringing her joy. She and the other women of Hull House frequently referred to themselves as "sisters," and they depended utterly on each other. Addams's closest relationship was with Mary Rozet Smith, with whom she spent most of her vacations and of whom Addams carried a painting when she traveled. In a letter to Smith, Addams wrote of her gratitude for their connection: "I feel as if we had come into a healing domesticity which we never had before, as if it were the best affection offered me."[9] Addams depended upon her friends and her companion, drawing strength and joy from her relationships.

Concerned people who are struggling for climate justice need friends and companions like Addams's. We need to seek balanced and sustainable joy, as Woolman did. We need to enjoy good food and good jokes, as King, Day, and Chavez did. The challenges we face are real, the injustices are grim, and the future is uncertain. This makes it all the more important that we learn to enjoy our lives, here and now.

## DIVERSE PATHS

The five witnesses considered in this book enjoyed life in different ways, which is appropriate because they also did everything else in different ways. Each blazed a particular path in response to particular circumstances. In the lives recounted here, there was no single style of resistance against structural violence but instead a series of very different choices, some of which are in significant tension with one other.

John Woolman's witness against slavery involved profound personal transformation; he sought to cleanse his life as much as possible from the trappings and privilege that a slave-holding society offered to white men. He stopped wearing dyed clothes and stopped expecting free labor as a witness against the violence of his society. Jane Addams was faced with an argument that she should live a similarly austere life when she visited Leo Tolstoy, who told her that she could only truly help the poor if she became poor herself. She considered his suggestion but ultimately rejected it as impractical. Addams continued to live as a wealthy woman who educated and empowered others while working toward a society where no one would be poor.

Addams's success came largely from her pragmatism and her willingness to compromise. She was a strict pacifist but was willing to sign on to a political party's platform that included war expenditures because she sought the best outcome possible in any given situation. By contrast, Dorothy Day was a woman of unshaking principle who drew moral authority from the consistency of her witness. So she refused to cooperate with a government that used violence, she abstained from paying taxes and voting, and she vocally opposed all preparations for war throughout her life, without compromise.

Day's movement was effective because it was so specifically located. She established a residential community at the Catholic Worker house in New York City, from which she fought hunger by feeding people, spoke out for justice by writing about their plight, and formed a movement that still considers the New York house its heart. By contrast, Martin Luther King Jr. ran a movement that regularly changed location as circumstances changed and was defined not by where he lived but by where he found injustice. He made homes in Montgomery, Atlanta, and, briefly, Chicago, but his protests took place wherever he believed he could draw national attention to the evils of segregation—Albany, Birmingham, Saint Augustine, Selma, Chicago, and Memphis. King could have established a single community that modeled justice and peace, but instead he chose to take his protest to the streets of whatever city could advance his cause.

King's protests were defined by his speeches, and he remains known today for his powerful voice, the passion of his words, and the soaring imagery he conjured. He was a preacher from a family and a culture of powerful preaching. By contrast, Cesar Chavez was a quiet speaker who made straightforward arguments in a monotone, with few flourishes. Audiences tended to listen closely because of Chavez's personal virtue and experience rather than his delivery. Like King, Chavez was charismatic and intelligent. But Chavez expressed himself in a much quieter, less oratorically dynamic way.

One could describe many more distinctions between these five witnesses. Some were devout, some were lax; some were Catholic, some were Protestant; some were rich, some were poor; two were women, three were men; three were white, one was African American, and one was Mexican American. The important point is that there is no single witness here. In these five lives, one finds myriad options for how to resist structural violence. We can learn from this that there is much we *could* do to work toward climate justice, and so there will be multiple options as we discern what we *should* do.

## UNFINISHED WORK

One final commonality is worth stressing. All five witnesses left behind unfinished struggles. John Woolman opposed legalized slavery and economic inequality, both of which expanded during his lifetime. Jane Addams and Dorothy Day devoted their lives to feeding the hungry and opposing war, and yet people were still hungry and wars were still fought as they died. Martin Luther King Jr. and Cesar Chavez both stood up against racism and the systemic injustices that came with it, and yet racism persists. None of these witnesses defeated structural violence.

This would not have surprised them. Day cautioned her readers against a sense of responsibility for solving the world's problems or seeking affirmation from any form of success. She asked, "And why must we see results? Our work is to sow. Another generation will be reaping the harvest."[10] King preached his last sermon with a profound hope for "the promised land" of a just future but also noted that "I may not get there with you."[11] They knew that the struggle would be unfinished, and they continued to struggle.

These witnesses were also incomplete because they were not aware of all the structural violence amid which they lived. Woolman, King, and Chavez paid too little attention to the oppression of women in their societies, in their movements, and in their own lives. Day was uninterested in criticisms of the Catholic Church's intolerance toward diverse expressions of sexuality. Despite her pragmatism, many people during Addams's life found her too politically extreme, while many who have written after her death have found her insufficiently radical. By twenty-first-century standards, Woolman was paternalistic, working to help slaves and Native Americans without first seeking to understand or empower them on their own terms. All five witnesses were imperfect, fallible people.

This would not have surprised them. Woolman, for example, has never had a critic harsher than himself. He writes in his journal that he committed to giving up dyed clothing only after being overwhelmed by "sadness" at his moral failings, "a sense of my own wretchedness" that drove him to "mourning."[12] Addams, similarly, devoted much ink to her mistakes. Her most famous book, *Twenty Years at Hull House,* intentionally included "the stress and storm" of her work, and she criticized herself for her naive ideas about her neighbors, her rude treatment of visitors, and the moral dilemmas she could not solve.[13] All these witnesses knew that they had limits.

Chavez, King, Day, Addams, and Woolman were imperfect, and their work was unfinished. These facts would be discouraging—if one believed that they should have led blameless lives that achieved absolute victories. But these facts can, instead, be encouraging if one takes them as a sign that these five witnesses were human beings called to do their best in the world they found with the tools they had available. All of them succeeded admirably at that task. All of them finished that work.

To see these witnesses in this way also helps us to understand our own task in the twenty-first century. Concerned people should not hope to end the injustice, inequalities, and destruction of climate change. Climate change—like war, racism, and inequality—is a wicked problem. It will not end in our lifetimes. We can make progress against structural violence and seek to alleviate the suffering it causes, but we will not finish these projects. We can work to plant seeds, but we should not expect to reap a complete harvest. We can hope to glimpse the mountaintop, but we should not expect to live in the promised land.

We must also accept that we will make mistakes. Indeed, we are making mistakes right now. We are limited and imperfect, just as these five witnesses were. Even as we work toward climate justice, we are surely making assumptions that future generations will look upon harshly. As we consider what more we can do, we are surely ignoring voices that we should be hearing. When we struggle with our moral finitude, we are surely missing things we could be doing to make the world more just. But this need not make us hopeless, if we enjoy the good company in which it puts us.

We are flawed—just like Woolman, Addams, Day, King, and Chavez. Like them, we should strive to do our best. And like them, we should accept that our best will never be perfect.

Climate change causes real pain and injustice in the world. This book has mentioned the plights of people who suffer from rising seas in Bangladesh, droughts in Sudan, encroaching homelessness Kivalina, and food shortages in Bolivia. As climate activists, we have the power to help these people, to alleviate some of this suffering. We can make the world more just. We can also continue to expand awareness of suffering in the non-human world, helping people learn to care about dwindling polar bear habitats, expanding kudzu, dying forests, and shrinking glaciers. We can do a great deal. But we cannot end all suffering and injustice. Realizing this will free us to do what we can. Our task is not to save the world but to resist structural violence as we encounter it, doing our best with the tools we have in the world as we find it.

## WITNESS OF RESISTANCE

So, as concerned people, what can we do about the injustice of climate change in the twenty-first century? We can work to enjoy a life of resistance; we can choose a path forward secure in the knowledge that there is no single, correct choice; and we can accept that our efforts will be limited and imperfect.

We can learn to see climate change as a form of structural violence linked to other forms of structural violence—like patriarchy, racism, classism, ableism, and heterosexism. We can take up the legacy of nonviolent resistance against such structural violence in order to make the world more just.

We can attend to the witnesses of five US Christians who made their world better through their commitments to nonviolence. Like John Woolman, we can seek to purify ourselves from the trappings of violence as much as we are able. Like Jane Addams, we can work for changes in our neighborhoods, our nation, and the global community. Like Dorothy Day, we can find faith enough to love the world and our neighbors who suffer from economic injustice. Like Martin Luther King Jr., we can learn what it means to have a realistic and radical hope for the future. And like Cesar

Chavez, we can make sacrifices for those less fortunate than ourselves and ask other privileged people to do the same. Finally, we can continue to look for more witnesses from throughout history and from our own time who inspire and instruct us.

So what can we do? We can work against the violence of climate change. We can work to become witnesses, leaving a legacy of resistance from which future generations can learn.

## NOTES

1. Mahan, *Forgetting Ourselves on Purpose*, 11.
2. Levy and Chavez, *Cesar Chavez*, 196.
3. Chavez, *Words of Cesar Chavez*, 80.
4. Quoted by Troester, *Voices from the Catholic Worker*, 72.
5. Quoted by Roberts, *Day and the Catholic Worker*, 75.
6. Burns, *To the Mountaintop*, 447–49.
7. In his book on the common legacy of Martin Luther King and Malcolm X, James Cone notes that they shared a willingness to link anger at injustice with humor at the ironies of an unjust world. He notes: "To fight for life is to experience the joy of life. To laugh, to have fun, is to bear witness to life against death. Freedom fighters are fun-loving people. Therefore, let us laugh, let us shout for joy, not as an indication that we are no longer angry but rather as a sign that we have just begun to fight." Cone, *Martin & Malcolm & America*, 309.
8. Woolman, *Journal and Major Essays*, 29.
9. Quoted by Joslin, *Jane Addams*.
10. Day, *By Little and By Little*, 92.
11. King, *Testament of Hope*, 263.
12. Woolman, *Journal and Major Essays*, 119.
13. Addams, *Twenty Years at Hull House*, 42.

# BIBLIOGRAPHY

Ackerman, Peter, and Jack DuVall. *A Force More Powerful: A Century of Nonviolent Conflict*. New York: St. Martin's Press, 2000.

Ackerman, Peter, and Christopher Kruegler. *Strategic Nonviolent Conflict: The Dynamics of People Power in the Twentieth Century*. Westport, CT: Praeger, 1994.

Addams, Jane. *Democracy and Social Ethics*. Bedford, MA: Applewood Books, 2002.

———. *The Jane Addams Reader*, edited by Jean Bethke Elshtain. New York: Basic Books, 2002.

———. *Jane Addams's Writings on Peace*, edited by Judy D. Whipps and Marilyn Fischer. Bristol: Thoemmes Press, 2003.

———. *Newer Ideals of Peace*. Urbana: University of Illinois Press, 2007.

———. *The Second Twenty Years at Hull-House: With a Record of a Growing World Consciousness*. New York: Macmillan, 1930.

———. *The Spirit of Youth and the City Streets*. New York: Macmillan, 1909.

———. *Twenty Years at Hull-House: With Autobiographical Notes*. Boston: Bedford / St. Martin's Press, 1999.

Adler, Jonathan H., ed. *Ecology, Liberty & Property: A Free Market Environmental Reader*. Washington, DC: Competitive Enterprise Institute, 2000.

Aiken, Scott F. "The Significance of Al Gore's Purported Hypocrisy." *Environmental Ethics* 31, no. 1 (2009): 111–12.

Allen, Joseph L. *War: A Primer for Christians*. Nashville: Abingdon Press, 1991.

Allman, Mark. *Who Would Jesus Kill? War, Peace, and the Christian Tradition*. Winona, MN: Saint Mary's Press, 2008.

Andolsen, Barbara Hilkert. "Agape in Feminist Ethics." In *Feminist Theological Ethics: A Reader*, edited by Lois Daly. Louisville: Westminster John Knox, 1994.

Baldwin, Lewis V. *There Is a Balm in Gilead: The Cultural Roots of Martin Luther King Jr.* Minneapolis: Fortress Press, 1991.

———. *To Make the Wounded Whole: The Cultural Legacy of Martin Luther King Jr.* Minneapolis: Fortress Press, 1992.

Bales, Kevin. *Disposable People: New Slavery in the Global Economy*. Berkeley: University of California Press, 2012.

Bauman, Whitney. *Religion and Ecology: Developing a Planetary Ethic*. New York: Columbia University Press, 2014.

Bauman, Whitney, Richard Bohannon, and Kevin J. O'Brien. *Inherited Land: The Changing Grounds of Religion and Ecology*. Eugene, OR: Pickwick, 2011.

Birkel, Michael. *Mysticism and Activism: Learning From John Woolman*. Melbourne Beach, FL: Southeastern Yearly Meeting, 2002.

———. *A Near Sympathy: The Timeless Quaker Wisdom of John Woolman*. Richmond, IN: Friends United Press, 2003.

Birt, Robert E., ed. *The Liberatory Thought of Martin Luther King Jr.: Critical Essays on the Philosopher King*. Lanham, MD: Lexington Books, 2012.

Blanchard, Kathryn D'Arcy, and Kevin J. O'Brien. *An Introduction to Christian Environmentalism: Ecology, Virtue, and Ethics*. Waco, TX: Baylor University Press, 2014.

———. "Prophets Meet Profits: What Christian Ecological Ethics Can Learn from Free Market Environmentalism." *Journal of the Society of Christian Ethics* 33, no. 2 (2014): 103–23.

Boserup, Anders, and Andrew Mack. *War Without Weapons: Non-Violence in National Defense*. New York: Schocken Books, 1975.

Bradley, Bethany A., David S. Wilcove, and Michael Oppenheimer. "Climate Change Increases Risk of Plant Invasion in the Eastern United States." *Biological Invasions* 12, no. 6 (2010): 1855–72.

Branch, Taylor. *At Canaan's Edge: America in the King Years, 1965–68*. New York: Simon & Schuster, 2006.

———. *Parting the Waters: America in the King Years, 1954–63*. New York: Simon & Schuster, 1988.

———. *Pillar of Fire: America in the King Years, 1963–65*. New York: Simon & Schuster, 1998.

Brimlow, Robert W. *What About Hitler? Wrestling with Jesus's Call to Nonviolence in an Evil World*. Grand Rapids: Brazos Press, 2006.

British Royal Society. *Geoengineering the Climate: Science, Governance, and Uncertainty*. London: British Royal Society, 2009.

Brown, Victoria. *The Education of Jane Addams*. Philadelphia: University of Pennsylvania Press, 2004.

———. "Sermon of the Deed: Jane Addams's Spiritual Evolution." In *Jane Addams and the Practice of Democracy*, edited by Marilyn Fischer, Carol Nackenoff, and Wendy E. Chmielewski. Urbana: University of Illinois Press, 2009.

Burkett, Maxine. "Climate Reparations." *Melbourne Journal of International Law* 10 (2009): 509–42.

Burns, Stewart. *To the Mountaintop: Martin Luther King Jr.'s Sacred Mission to Save America, 1955–1968*. San Francisco: HarperSanFrancisco, 2004.

Burns, William C. G., and Andrew L. Strauss, eds. *Climate Change Geoengineering: Philosophical Perspectives, Legal Issues, and Governance Frameworks.* New York: Cambridge University Press, 2013.

Burrow, Rufus, Jr. *Extremist for Love: Martin Luther King Jr.: Man of Ideas and Non-violent Social Action.* Minneapolis: Fortress Press, 2014.

———. *God and Human Dignity: The Personalism, Theology, and Ethics of Martin Luther King Jr.* Notre Dame, IN: University of Notre Dame Press, 2006.

Cahill, Lisa Sowle. *Love Your Enemies: Discipleship, Pacifism, and Just War Theory.* Minneapolis: Fortress Press, 1994.

Callicott, J. Baird. *Beyond the Land Ethic: More Essays in Environmental Philosophy.* Albany: State University of New York Press, 1999.

Carey, Brycchan, and Geoffrey Gilbert Plank. *Quakers and Abolition.* Champaign: University of Illinois Press, 2014.

Carson, Rachel. *Silent Spring.* New York: Houghton Mifflin, 1994.

Chappell, David L. *A Stone of Hope: Prophetic Religion and the Death of Jim Crow.* Chapel Hill: University of North Carolina Press, 2004.

Chavez, Cesar. *An Organizer's Tale: Speeches,* edited by Ilan Stavans. New York: Penguin, 2008.

———. *The Words of Cesar Chavez,* edited by Richard J. Jensen and John C. Hammer-back. College Station: Texas A&M University Press, 2002.

Chernus, Ira. *American Nonviolence: The History of an Idea.* Maryknoll, NY: Orbis Books, 2004.

Clingerman, Forrest. "Between Babel and Pelagius: Religion, Theology, and Geo-engineering." In *Engineering the Climate: The Ethics of Solar Radiation Management,* edited by Christopher J. Preston. Lanham, MD: Lexington Books, 2012.

Clingerman, Forrest, and Kevin J. O'Brien. "Playing God: Why Religion Belongs in the Climate Engineering Debate." *Bulletin of the Atomic Scientists* 70, no. 3 (2014): 27–37.

———. *Theological and Ethical Perspectives on Climate Engineering: Calming the Storm.* Lanham, MD: Lexington Books, 2016.

Coles, Robert. *Dorothy Day: A Radical Devotion.* Reading, MA: Addison-Wesley, 1987.

Collin, Robin Morris, and Robert Collin. "Environmental Reparations." In *The Quest for Environmental Justice: Human Rights and the Politics of Pollution,* edited by Robert D. Bullard. San Francisco: Sierra Club Books, 2005.

Cone, James H. *The Cross and the Lynching Tree.* Maryknoll, NY: Orbis Books, 2011.

———. *Martin & Malcolm & America: A Dream or a Nightmare.* Maryknoll, NY: Orbis Books, 1991.

———. "Whose Earth Is It, Anyway?" In *Earth Habitat: Eco-Injustice and the Church's Response,* edited by Dieter T. Hessel and Larry L. Rasmussen. Minneapolis: Fortress Press, 2001.

Conference of the Parties. "The Paris Agreement." December 11, 2015. www
.google.com/url?sa=t&rct=j&q=&esrc=s&source=web&cd=1&ved=
0ahUKEwj5gJPMm_XJAhVKyGMKHfrzC-YQFggdMAA&url=http%3A
%2F%2Funfccc.int%2Fresource%2Fdocs%2F2015%2Fcop21%2Feng
%2Fl09r01.pdf&usg=AFQjCNEzbDnhBToMrBraS7NQ_6bpUrOzlQ&sig2=
jo5yyBih9npjMGiRgXPRzA.

Conradie, Ernst M. *The Church and Climate Change*. Pietermaitzburg, South Africa:
Cluster, 2008.

Cornell, Thomas C., Robert Ellsberg, and Jim Forest. *A Penny a Copy: Readings from
the Catholic Worker*. Maryknoll, NY: Orbis Books, 1995.

Cortright, David. *Gandhi and Beyond: Nonviolence for a New Political Age*. Boulder,
CO: Paradigm, 2009.

Cox, Louis, and Ruah Swennerfelt. *Walking in the Light: John Woolman's Witness for
Living in Right Relationship with All of Creation*. Charlotte, VT: Peace for Earth,
2007.

Dalton, Frederick John. *The Moral Vision of Cesar Chavez*. Maryknoll, NY: Orbis
Books, 2003.

Davis, Vernie. "John Woolman and Structural Violence: Model for Analysis and
Social Change." In *The Tendering Presence: Essays on John Woolman*, edited by
Mike Heller. Wallingford, PA: Pendle Hill, 2003.

Dawson, Clanton C. W., Jr. "The Concept of Hope in the Thinking of Dr. Martin
Luther King Jr." In *The Liberatory Thought of Martin Luther King Jr.: Critical
Essays on the Philosopher King*, edited by Robert E. Birt. Lanham, MD: Lexing-
ton Books, 2012.

Day, Dorothy. *By Little and By Little: The Selected Writings of Dorothy Day*, edited by
Robert Ellsberg. New York: Alfred A. Knopf, 1983.

———. *From Union Square to Rome*. Maryknoll, NY: Orbis Books, 2006.

———. *House of Hospitality*. New York: Sheed & Ward, 1939.

———. *The Long Loneliness: The Autobiography of Dorothy Day*. San Francisco:
Harper & Row, 1981.

———. *On Pilgrimage*. Grand Rapids: William B. Eerdmans, 1999.

———. *On Pilgrimage: The Sixties*. New York: Curtis Books, 1972.

Del Castillo, Richard Griswold, and Richard A. Garcia. *Cesar Chavez: A Triumph of
Spirit*. Norman: University of Oklahoma Press, 1995.

Deming, Barbara. *We Are All Part of One Another: A Barbara Deming Reader*. Philadelphia: New Society, 1984.

Derocher, Andrew E., Jon Aars, Steven C. Amstrup, Amy Cutting, Nick J. Lunn,
Péter K. Molnár, Martyn E. Obbard, Ian Stirling, Gregory W. Theimann, Dag
Vongraven, Øystein Wiig, and Geoffrey York. "Rapid Ecosystem Change and
Polar Bear Conservation." *Conservation Letters* 6, no. 5 (2013): 368–75.

Dostoyevsky, Fyodor. *The Brothers Karamazov*. New York: Macmillan, 1919.

Eaton, Marie. "Navigating Anger, Fear, Grief, and Despair." In *Contemplative Approaches to Sustainability in Higher Education: Theory and Practice*, edited by Marie Eaton, Holly J. Hughes, and Jean MacGregor. New York: Routledge, 2017.

Elshtain, Jean Bethke. *Jane Addams and the Dream of American Democracy: A Life*. New York: Basic Books, 2002.

Etulain, Richard W. *Cesar Chavez: A Brief Biography with Documents*. New York: Palgrave, 2002.

Ferriss, Susan, Ricardo Sandoval, and Diana Hembree. *The Fight in the Fields: Cesar Chavez and the Farmworkers Movement*. New York: Harcourt Brace, 1997.

Fischer, Marilyn. *On Addams*. Belmont, CA: Thomson Wadsworth, 2004.

Fischer, Marilyn, Carol Nackenoff, and Wendy E. Chmielewski, eds. *Jane Addams and the Practice of Democracy*. Urbana: University of Illinois Press, 2009.

Forest, Jim. *All Is Grace: A Biography of Dorothy Day*. Maryknoll, NY: Orbis Books, 2011.

Francis, John. *Planetwalker*. Washington, DC: National Geographic, 2008.

Franklin, Robert Michael, Jr. "An Ethic of Hope: The Moral Thought of Martin Luther King Jr." *Union Seminary Quarterly Review* 40, no. 4 (1985): 41–51.

Galtung, Johan. "Violence, Peace, and Peace Research." *Journal of Peace Research* 6, no. 3 (1969): 167–91.

Gandhi, Mohandas K. *All Men Are Brothers: Autobiographical Reflections*. New York: Continuum, 2005.

———. *Autobiography: The Story of My Experiments with Truth*. New York: Dover, 1983.

———. *The Essential Gandhi: An Anthology*, edited by Louis Fischer. New York: Random House, 1962.

Ganz, Marshal. *Why David Sometimes Wins: Leadership, Organization, and Strategy in the California Farm Worker Movement*. New York: Oxford University Press, 2009.

Garrow, David J. "The Intellectual Development of Martin Luther King Jr.: Influences and Commentaries." *Union Seminary Quarterly Review* 40, no. 4 (1985): 5–20.

———. "Martin Luther King Jr. and the Spirit of Leadership." *Journal of American History* 74, no. 2 (1987): 438–47.

Gore, Albert. *An Inconvenient Truth: The Planetary Emergency of Global Warming and What We Can Do About It*. Emmaus, PA: Rodale Press, 2006.

Gottlieb, Roger S. *A Greener Faith: Religious Environmentalism and Our Planet's Future*. New York: Oxford University Press, 2006.

Gough, Ian, Saamah Abdallah, Victoria Johnson, Josh Ryan-Collins, and Cindy Smith. "The Distribution of Total Greenhouse Gas Emissions by Households in the UK, and Some Implications for Social Policy." Centre for Analysis of Social

Exclusion, case 152, 2011. http://eprints.lse.ac.uk/36562/1/The%20distribution
%20of%20total%20greenhouse%20gas%20emissions%20by%20households%20in
%20the%20UK,%20and%20some%20implications%20for%20social%20policy
(revised%20for%20LSERO).pdf.

Green, Judith M. "Social Democracy, Cosmopolitan Hospitality, and Interciviliza-
tional Peace: Lessons from Jane Addams." In *Feminist Interpretations of Jane
Addams*, edited by Maurice Hamington. University Park: Pennsylvania State
University Press, 2010.

Grey, Mary C. *Sacred Longings: The Ecological Spirit and Global Culture*. Minneapolis: Fortress Press, 2004.

Grim, John, and Mary Evelyn Tucker. *Ecology and Religion*. Washington, DC: Island
Press, 2014.

Gunster, Shane. "Self-Interest, Sacrifice, and Climate Change: (Re)Framing the British Columbia Carbon Tax." In *The Environmental Politics of Sacrifice*, edited by
Michael Maniates and John M. Meyer. Cambridge, MA: MIT Press, 2010.

Hallie, Philip Paul. *Lest Innocent Blood Be Shed: The Story of the Village of Le Chambon and How Goodness Happened There*. New York: Harper Perennial, 1994.

Hamilton, Clive. *Earthmasters: The Dawn of the Age of Climate Engineering*. New
Haven, CT: Yale University Press, 2013.

——. *Requiem for a Species: Why We Resist the Truth About Climate Change*. Washington, DC: Earthscan, 2010.

Hamington, Maurice. *Feminist Interpretations of Jane Addams*. University Park:
Pennsylvania State University Press, 2010.

——. *The Social Philosophy of Jane Addams*. Urbana: University of Illinois Press,
2009.

——. "Two Leaders, Two Utopias: Jane Addams and Dorothy Day." *NWSA Journal*
19, no. 2 (2007): 159–86.

Hansen, J., M. Sato, P. Kharecha, D. Beerling, R. Berner, V. Masson-Delmotte,
M. Pagani, M. Raymo, D. L. Royer, and J. C. Zachos. "Target Atmospheric $CO_2$:
Where Should Humanity Aim?" *Open Atmospheric Science Journal* 2 (2008):
217–31.

Harding, Vincent. *Martin Luther King, the Inconvenient Hero*. Maryknoll, NY: Orbis
Books, 1996.

Harris, Melanie L. "Ecowomanism: An Introduction." *Worldviews* 20 (2016): 5–14.

Hartman, Laura M. *The Christian Consumer: Living Faithfully in a Fragile World*.
New York: Oxford University Press, 2011.

Harvey, Jennifer. "Dangerous 'Goods': Seven Reasons Creation Care Movements
Must Advocate Reparations." In *Buffalo Shout, Salmon Cry: Conversations on
Creation, Land Justice, and Life Together*, edited by Steve Heinrichs. Waterloo,
Ontario: Herald Press, 2013.

Hayes, Floyd W., III. "Hope and Disappointment in Martin Luther King Jr.'s Political Theology: Eclipse of the Liberal Spirit." In *The Liberatory Thought of Martin Luther King Jr.: Critical Essays on the Philosopher King*, edited by Robert E. Birt. Lexington Books: Lanham, MD, 2012.

Heller, Mike, ed. *The Tendering Presence: Essays on John Woolman.* Wallingford, PA: Pendle Hill, 2003.

Henning, Brian G. *Riders in the Storm: Ethics in an Age of Climate Change.* Winona, MN: Anselm Academic, 2015.

"Hindu Declaration on Climate Change." *Hinduism Today*, April 1, 2010.

Hoekema, David A. "A Practical Christian Pacifism." *Christian Century*, October 22, 1986, 917–19.

Hribar, Paul Anthony. "The Social Fasts of Cesar Chavez." PhD dissertation, University of Southern California, 1978.

Hulme, Mike. *Can Science Fix Climate Change? A Case Against Climate Engineering.* Boston: Polity, 2014.

———. *Why We Disagree About Climate Change: Understanding Controversy, Inaction and Opportunity.* New York: Cambridge University Press, 2009.

IPCC (Intergovernmental Panel on Climate Change). "Summary for Policymakers." In *Climate Change 2013: The Physical Science Basis—Contribution of Working Group I to the Fifth Assessment Report of the Intergovernmental Panel on Climate Change*, edited by T. F. Stocker, D. Qin, G.-K. Plattner, M. Tignor, S. K. Allen, J. Boschung, A. Nauels, Y. Xia, V. Bex, and P. M. Midgley. New York: Cambridge University Press, 2013.

———. "Summary for Policymakers." In *Climate Change 2014: Impacts, Adaptation, and Vulnerability, Part A—Global and Sectoral Aspects; Contribution of Working Group II to the Fifth Assessment Report of the Intergovernmental Panel on Climate Change*, edited by C. B. Field, V. R. Barros, D. J. Dokken, K. J. Mach, M. D. Mastrandrea, T. E. Bilir, M. Chatterjee, K. L. Ebi, Y. O. Estrada, R. C. Genova, B. Girma, E. S. Kissel, A. N. Levy, S. MacCracken, P. R. Mastrandrea, and L. L. White. New York: Cambridge University Press, 2014.

———. "Summary for Policymakers." In *Climate Change 2014: Mitigation of Climate Change—Contribution of Working Group III to the Fifth Assessment Report of the Intergovernmental Panel on Climate Change*, edited by O. Edenhofer, R. Pichs-Madruga, Y. Sokona, E. Farahani, S. Kadner, K. Seyboth, A. Adler, I. Baum, S. Brunner, P. Eickemeier, B. Kriemann, J. Savolainen, S. Schlömer, C. von Stechow, T. Zwickel, and J. C. Minx. New York: Cambridge University Press, 2014.

James, William. "The Moral Equivalent of War." In *William James: The Essential Writings*, edited by Bruce W. Wilshire. Albany: State University of New York Press, 1984.

Jenkins, Willis. *The Future of Ethics: Sustainability, Social Justice, and Religious Creativity.* Washington, DC: Georgetown University Press, 2013.

Johnston, Lucas F. *Religion and Sustainability: Social Movements and the Politics of the Environment.* Bristol, CT: Equinox, 2013.

Joslin, Katherine. *Jane Addams: A Writer's Life.* Urbana: University of Illinois Press, 2004.

Juhnke, James C., and Carol Hunter. *The Missing Peace: The Search for Nonviolent Alternatives in United States History.* Kitchener, Ontario: Pandora Press, 2001.

Keith, David W. *A Case for Climate Engineering.* Cambridge, MA: MIT Press, 2013.

Kim, Grace Ji-Sun, ed. *Making Peace with the Earth: Action and Advocacy for Climate Justice.* Geneva: World Council of Churches, 2016.

King, Martin Luther, Jr. *The Autobiography of Martin Luther King, Jr.*, edited by Clayborne Carson. New York: Warner Books, 1998.

———. *"In a Single Garment of Destiny": A Global Vision of Justice*, edited by Lewis V. Baldwin. Boston: Beacon Press, 2012.

———. *Strength to Love.* Philadelphia: Fortress Press, 1981.

———. *Stride toward Freedom: The Montgomery Story.* New York: Harper Brothers, 1958.

———. *A Testament of Hope: The Essential Writings of Martin Luther King, Jr.*, edited by James Melvin Washington. San Francisco: HarperCollins, 1986.

———. *The Trumpet of Conscience.* New York: Harper & Row, 1968.

———. *Where Do We Go from Here—Chaos or Community?* Boston: Beacon Press, 1967.

———. *Why We Can't Wait.* New York: Penguin, 1964.

Klein, Naomi. *This Changes Everything: Capitalism vs. the Climate.* New York: Simon & Schuster, 2014.

Klejment, Anne. "The Spirituality of Dorothy Day's Pacifism." *US Catholic Historian* 27, no. 2 (2009): 1–24.

Klejment, Anne, and Nancy L. Roberts, eds. *American Catholic Pacifism: The Influence of Dorothy Day and the Catholic Worker Movement.* Westport, CT: Praeger, 1996.

Knight, Louise W. *Citizen: Jane Addams and the Struggle for Democracy.* Chicago: University of Chicago Press, 2005.

———. "Garbage and Democracy." *Journal of Community Practice* 14, no. 3 (2006): 7–27.

———. *Jane Addams: Spirit in Action.* New York: Norton, 2010.

———. "Love on Halsted Street: A Contemplation on Jane Addams." In *Feminist Interpretations of Jane Addams*, edited by Maurice Hamington. University Park: Pennsylvania State University Press, 2010.

Kraybill, Donald B., Steven M. Nolt, and David Weaver-Zercher. *Amish Grace: How Forgiveness Transcended Tragedy.* San Francisco: Jossey-Bass, 2007.

———. *The Amish Way: Patient Faith in a Perilous World.* San Francisco: Jossey-Bass, 2010.

Kurlansky, Mark. *Nonviolence: The History of a Dangerous Idea*. New York: Modern Library, 2006.

León, Luis D. *The Political Spirituality of Cesar Chavez: Crossing Religious Borders*. Berkeley: University of California Press, 2015.

Leopold, Aldo. *A Sand County Almanac, with Essays on Conservation from Round River*. New York: Ballantine Books, 1966.

Lévinas, Emmanuel. *Difficult Freedom: Essays on Judaism*. Baltimore: Johns Hopkins University Press, 1990.

Levy, Jacques E., and Cesar Chavez. *Cesar Chavez: Autobiography of La Causa*. New York: Norton, 1975.

Lewis, Avi (director), and Naomi Klein (narrator). *This Changes Everything*. Film, streaming. New York: FilmBuff Studio, 2015.

Llanos, Angélica Navarro. "Climate Debt: The Basis of a Fair and Effective Solution to Climate Change." Technical Briefing on Historical Responsibility, United Nations Framework Convention on Climate Change, June 4, 2009. https://unfccc .int/files/meetings/ad_hoc_working_groups/lca/application/pdf/4_bolivia.pdf.

Lomborg, Bjørn. *Cool It: The Skeptical Environmentalist's Guide to Global Warming*. New York: Vintage Books, 2008.

———. *The Skeptical Environmentalist: Measuring the Real State of the World*. New York: Cambridge University Press, 2001.

———, ed. *Smart Solutions to Climate Change: Comparing Costs and Benefits*. New York: Cambridge University Press, 2010.

Long, Michael G. *Christian Peace and Nonviolence: A Documentary History*. Maryknoll, NY: Orbis Books, 2011.

Loy, David. *Money, Sex, War, Karma: Notes for a Buddhist Revolution*. Boston: Wisdom Publications, 2008.

———. "The Religion of the Market." *Journal of the American Academy of Religion* 65, no. 2 (1997): 275–90.

Mahan, Brian. *Forgetting Ourselves on Purpose: Vocation and the Ethics of Ambition*. San Francisco: Jossey-Bass, 2002.

Mandela, Nelson. *Long Walk to Freedom: The Autobiography of Nelson Mandela*. Boston: Little, Brown, 1994.

Maniates, Michael, and John M. Meyer. *The Environmental Politics of Sacrifice*. Cambridge, MA: MIT Press, 2010.

Marshall, Ellen Ott. "Practicing Imagination." In *Choosing Peace through Daily Practices*, edited by Ellen Ott Marshall. Cleveland: Pilgrim Press, 2005.

Marshall, George. *Don't Even Think About It: Why Our Brains Are Wired to Ignore Climate Change*. New York: Bloomsbury, 2014.

Martin-Schramm, James B. *Climate Justice: Ethics, Energy, and Public Policy*. Minneapolis: Fortress Press, 2010.

Matthew, Richard. "Climate Change and Human Security." In *Climate Change: What It Means for Us, Our Children, and Our Grandchildren*, edited by Joseph F. C. DiMento and Pamela Doughman. Cambridge, MA: MIT Press, 2014.

Matthiessen, Peter. *Sal Si Puedes: Cesar Chavez and the New American Revolution*. New York: Random House, 1969.

McFague, Sallie. *Blessed Are the Consumers: Climate Change and the Practice of Restraint*. Minneapolis: Fortress Press, 2013.

———. *Life Abundant: Rethinking Theology and Economy for a Planet in Peril*. Minneapolis: Fortress Press, 2001.

———. *A New Climate for Theology: God, the World, and Global Warming*. Minneapolis: Fortress Press, 2008.

McKibben, Bill. *Eaarth: Making a Life on a Tough New Planet*. New York: Times Books, 2010.

———. *The End of Nature*. New York: Random House, 1989.

———. "Global Warming's Terrifying New Math." *Rolling Stone*, August 2, 2012.

———. "How the Active Many Can Overcome the Ruthless Few." *The Nation*, December 19–26, 2016.

———. "A World at War." *The New Republic*, August 15, 2016.

Mercedes, Anna. *Power for: Feminism and Christ's Self-Giving*. New York: T&T Clark, 2011.

Miller, William D. *Dorothy Day: A Biography*. San Francisco: Harper & Row, 1982.

Minteer, Ben A., and Robert E. Manning. "Pragmatism in Environmental Ethics: Democracy, Pluralism, and the Management of Nature." *Environmental Ethics* 21, no. 2 (1999): 191–207.

Moe-Lobeda, Cynthia D. *Resisting Structural Evil: Love as Ecological-Economic Vocation*. Minneapolis: Fortress Press, 2013.

Moffett, Stephen R. Lloyd. "Holy Activist, Secular Saint: Religion and the Social Activism of Cesar Chavez." In *Mexican American Religions: Spirituality, Activism, and Culture*, edited by Gastón Espinosa and Mario T. Garcia. Durham, NC: Duke University Press, 2008.

Monbiot, George. *Heat: How to Stop the Planet from Burning*. Cambridge, MA: South End Press, 2007.

Moore, Kathleen Dean, and Michael P. Nelson. *Moral Ground: Ethical Action for a Planet in Peril*. San Antonio: Trinity University Press, 2010.

Moses, Greg. *Revolution of Conscience: Martin Luther King Jr. and the Philosophy of Nonviolence*. New York: Guilford Press, 1997.

Moulton, Phillips. "John Woolman's Approach to Social Action—as Exemplified in Relation to Slavery." *Church History Studies* 35, no. 4 (1966): 399–410.

Myles, Anne G. "'Stranger Friend': John Woolman and Quaker Dissent." In *The Tendering Presence: Essays on John Woolman*, edited by Mike Heller. Wallingford, PA: Pendle Hill, 2003.

Nash, Gary B., and Jean R. Soderlund. *Freedom by Degrees: Emancipation in Pennsylvania and Its Aftermath*. New York: Oxford University Press, 1991.

Nash, James A. *Loving Nature: Ecological Integrity and Christian Responsibility*. Nashville: Abingdon Press, 1991.

Niebuhr, Reinhold. "Why the Church Is Not Pacifist: Christianity and Power Politics, 1940." In *The End of Illusions: Religious Leaders Confront Hitler's Gathering Storm*, edited by Joseph Loconte. Lanham, MD: Rowman & Littlefield, 2004.

Nixon, Rob. *Slow Violence and the Environmentalism of the Poor*. Cambridge, MA: Harvard University Press, 2011.

Norton, Bryan G. *Toward Unity among Environmentalists*. New York: Oxford University Press, 1991.

O'Brien, Kevin J. *An Ethics of Biodiversity: Christianity, Ecology, and the Variety of Life*. Washington, DC: Georgetown University Press, 2010.

———. "The 'War' Against Climate Change and Christian Eco-Justice: Ethical Implications of Martial Rhetoric." *Worldviews* 16 (2012): 135–53.

O'Connor, June. *The Moral Vision of Dorothy Day: A Feminist Perspective*. New York: Crossroad, 1991.

O'Gorman, Angie, ed. *The Universe Bends toward Justice: A Reader on Christian Nonviolence in the US*. Philadelphia: New Society, 1990.

Orr, David W. *The Nature of Design: Ecology, Culture, and Human Intention*. New York: Oxford University Press, 2002.

Oxfam America. *Like Machines in the Fields: Workers Without Rights in American Agriculture*. Boston: Oxfam America, 2004.

Parenti, Christian. *Tropic of Chaos: Climate Change and the New Geography of Violence*. New York: Nation Books, 2011.

Parker, Theodore. *The Collected Works of Theodore Parker*, edited by Frances Power Cobbe. London: Trübner, 1879.

Pawel, Miriam. *The Crusades of Cesar Chavez: A Biography*. New York: Bloomsbury Press, 2014.

———. *The Union of Their Dreams: Power, Hope, and Struggle in Cesar Chavez's Farm Worker Movement*. New York: Bloomsbury Press, 2009.

Peterson, Anna. "Ordinary and Extraordinary Sacrifices: Religion, Everyday Life, and Environmental Practice." In *The Environmental Politics of Sacrifice*, edited by Michael Maniates and John M. Meyer. Cambridge, MA: MIT Press, 2010.

Pickering, Jonathan, and Christian Barry. "On the Concept of Climate Debt: Its Moral and Political Value." *Critical Review of International Social and Political Philosophy* 15, no. 5 (2012): 667–85.

Piehl, Mel. *Breaking Bread: The Catholic Worker and the Origin of Catholic Radicalism in America*. Philadelphia: Temple University Press, 1982.

Pipher, Mary Bray. *The Green Boat: Reviving Ourselves in Our Capsized Culture*. New York: Riverhead Books, 2013.

Plank, Geoffrey Gilbert. *John Woolman's Path to the Peaceable Kingdom: A Quaker in the British Empire*. Philadelphia: University of Pennsylvania Press, 2012.

Pooley, Eric. *The Climate War: True Believers, Power Brokers, and the Eleventh-Hour Fight to Save the Earth*. New York: Hyperion, 2010.

Pope Francis. *Laudato Si': On Care for Our Common Home*. Encyclical Letter. Vatican City: The Vatican, 2015. http://w2.vatican.va/content/francesco/en/encyclicals /documents/papa-francesco_20150524_enciclica-laudato-si.html.

Prouty, Marco G. *Cesar Chavez, the Catholic Bishops, and the Farmworkers' Struggle for Social Justice*. Tucson: University of Arizona Press, 2006.

Rasmussen, Larry L. *Earth Community, Earth Ethics*. Maryknoll, NY: Orbis Books, 1996.

———. *Earth-Honoring Faith: Religious Ethics in a New Key*. New York: Oxford University Press, 2013.

———. "Environmental Racism and Environmental Justice: Moral Theory in the Making?" *Journal of the Society of Christian Ethics* 24, no. 1 (2004): 3–28.

Rawles, Kate. "A Copernican Revolution in Ethics." In *Moral Ground: Ethical Action for a Planet in Peril*, edited by Kathleen Dean Moore and Michael P. Nelson. San Antonio: Trinity University Press, 2010.

Riegle, Rosalie G. *Dorothy Day: Portraits by Those Who Knew Her*. Maryknoll, NY: Orbis Books, 2006.

Rittel, Horst J. W., and Melvin M. Webber. "Dilemmas in a General Theory of Planning." *Policy Sciences* 4, no. 2 (1973): 155–69.

Roberts, Nancy L. *Dorothy Day and the Catholic Worker*. Albany: State University of New York Press, 1984.

Robock, Alan. "20 Reasons Why Geoengineering May Be a Bad Idea." *Bulletin of the Atomic Scientists* 64, no. 2 (2008): 14–18.

Rodriguez, Richard. *Darling: A Spiritual Autobiography*. New York: Viking, 2013.

Saiving, Valerie. "The Human Situation: A Feminine View." In *Womanspirit Rising: A Feminist Reader in Religion*, edited by Carol P. Christ and Judith Plaskow. San Francisco: HarperCollins, 1992.

Sauvage, Pierre (director). *Weapons of the Spirit*. Film, DVD. Los Angeles: Le Chambon Foundation, 2007 (orig. released 1987).

Schneiderhan, Erik. *The Size of Others' Burdens: Barack Obama, Jane Addams, and the Politics of Helping Others*. Stanford, CA: Stanford University Press, 2015.

Schock, Kurt. *Unarmed Insurrections: People Power Movements in Nondemocracies*. Minneapolis: University of Minnesota Press, 2005.

Schweitzer, Jeff, and Giuseppe Notorbartolo-Di Sciar. *A New Moral Code*. Los Angeles: Jacquie Jordan, 2010.

Scruton, Roger. *How to Think Seriously About the Planet: The Case for an Environmental Conservatism*. New York: Oxford University Press, 2012.

Shanley, Brayton. *The Many Sides of Peace: Christian Nonviolence, the Contemplative Life, and Sustainable Living*. Eugene, OR: Resource Publications, 2013.

Sharp, Gene. "From Dictatorship to Democracy: A Conceptual Framework for Liberation." 2010. www.aeinstein.org/wp-content/uploads/2013/09/FDTD.pdf.

———. *The Politics of Nonviolent Action, Part One: Power and Struggle*. Boston: Extending Horizons Books, 1973.

———. *The Politics of Nonviolent Action, Part Two: The Methods of Nonviolent Action*. Boston: Extending Horizons Books, 1973.

Sharp, Gene, and Joshua Paulson. *Waging Nonviolent Struggle: 20th-Century Practice and 21st-Century Potential*. Boston: Extending Horizons Books, 2005.

Shaw, Randy. *Beyond the Fields: Cesar Chavez, the UFW, and the Struggle for Justice in the 21st Century*. Berkeley: University of California Press, 2008.

Shearer, Christine. *Kivalina: A Climate Change Story*. Chicago: Haymarket Books, 2011.

Shue, Henry. *Climate Justice: Vulnerability and Protection*. New York: Oxford Univesity Press, 2014.

Sicius, Francis. "Prophecy Faces Tradition: The Pacifist Debate During World War II." In *American Catholic Pacifism: The Influence of Dorothy Day and the Catholic Worker Movement*, edited by Anne Klejment and Nancy L. Roberts. Westport, CT: Praeger, 1996.

Sider, Ronald J. *Christ and Violence*. Scottdale, PA: Herald Press, 1979.

———. *Nonviolent Action: What Christian Ethics Demands but Most Christians Have Never Really Tried*. Grand Rapids: Brazos Press, 2015.

Sitkoff, Harvard. *King: Pilgrimage to the Mountaintop*. New York: Hill & Wang, 2008.

Slaughter, Thomas P. *The Beautiful Soul of John Woolman, Apostle of Abolition*. New York: Hill & Wang, 2008.

Smith-Christopher, Daniel L. *Subverting Hatred: The Challenge of Nonviolence in Religious Traditions*. Maryknoll, NY: Orbis Books, 2000.

Smith, Fred L. "A Risk–Risk Approach to Climate Change." *Ecology, Liberty & Property: A Free Market Environmental Reader*, edited by Jonathan H. Adler. Washington, DC: Competitive Enterprise Institute, 2000.

Soderlund, Jean R. *Quakers & Slavery: A Divided Spirit*. Princeton: Princeton University Press, 1985.

Stanley, John, David Loy, and Dorje Gyurme. *A Buddhist Response to the Climate Emergency*. Boston: Wisdom Publications, 2009.

Stebner, Eleanor J. "The Theology of Jane Addams: Religion 'Seeking Its Own Adjustment.'" In *Feminist Interpretations of Jane Addams*, edited by Maurice Hamington. University Park: Pennsylvania State University Press, 2010.

Stephenson, Wen. *What We're Fighting for Now Is Each Other: Dispatches from the Front Lines of Climate Justice*. Boston: Beacon Press, 2015.

Stirling, Ian, and Andrew E. Derocher. "Effects of Climate Warming on Polar Bears: A Review of the Evidence." *Global Change Biology* 18, no. 9 (2012): 2684–2706.

Stivers, Laura. *Disrupting Homelessness: Alternative Christian Approaches*. Minneapolis: Fortress Press, 2011.

Stone, Christopher D. *Earth and Other Ethics: The Case for Moral Pluralism*. New York: Harper & Row, 1987.

Taylor, Bron Raymond. *Dark Green Religion: Nature Spirituality and the Planetary Future*. Berkeley: University of California Press, 2010.

———. "From the Ground Up: Dark Green Religion and the Environmental Future." In *Ecology and the Environment*, edited by Donald K. Swearer. Cambridge, MA: Harvard University Press, 2009.

———. "On Sacred or Secular Ground? Callicott and Environmental Ethics." *Worldviews* 1 (1997): 99–111.

Thorn, William J., Phillip M. Runkel, and Susan Mountin. *Dorothy Day and the Catholic Worker Movement: Centenary Essays*. Milwaukee: Marquette University Press, 2001.

Townes, Emilie Maureen. *Womanist Ethics and the Cultural Production of Evil*. New York: Palgrave Macmillan, 2006.

Trocmé, André. *Jesus and the Nonviolent Revolution*. Maryknoll, NY: Orbis Books, 2004.

Troester, Rosalie Riegle. *Voices from the Catholic Worker*. Philadelphia: Temple University Press, 1993.

Trueblood, Elton. *The People Called Quakers*. New York: Harper & Row, 1966.

Tutu, Desmond. *God Has a Dream: A Vision of Hope for Our Time*. New York: Doubleday, 2004.

Vonnegut, Kurt. *Slaughterhouse-Five, or the Children's Crusade: A Duty-Dance with Death*. New York: Delacorte, 1969.

Watley, William D. *Roots of Resistance: The Nonviolent Ethic of Martin Luther King Jr.* Valley Forge, PA: Judson Press, 1985.

Watt, Alan J. *Farm Workers and the Churches: The Movement in California and Texas*. College Station: Texas A&M University Press, 2010.

White, Lynn, Jr. "The Historical Roots of Our Ecologic Crisis." *Science* 155, no. 3767 (1967): 1203–7.

Williams, Terry Tempest. "What Love Looks Like: A Conversation with Tim DeChristopher." *Orion Magazine*, January–February 2012.

Windle, Phyllis. "The Ecology of Grief." In *Ecopsychology: Restoring the Earth, Healing the Mind*, edited by Theodore Roszak, Mary E. Gomes, and Allen D. Kanner. San Francisco: Sierra Club Books, 1995.

Wink, Walter. *Engaging the Powers: Discernment and Resistance in a World of Domination*. Minneapolis: Fortress Press, 1992.

———. *The Powers That Be: Theology for a New Millennium.* New York: Doubleday, 1998.

Woolman, John. *John Woolman and the Affairs of Truth: The Journalist's Essays, Epistles, and Ephemera*, edited by James Proud. San Francisco: Inner Light Books, 2010.

———. *The Journal and Major Essays of John Woolman*, edited by Phillips P. Moulton. Richmond, IN: Friends United Press, 1971.

Yoder, John Howard. *The Politics of Jesus: Vicit Agnus Noster.* Grand Rapids: William B. Eerdmans, 1994.

Yordy, Laura. *Green Witness: Ecology, Ethics, and the Kingdom of God.* Eugene, OR: Cascade, 2010.

Zachariah, George. *Alternatives Unincorporated: Earth Ethics From the Grassroots.* London: Equinox, 2011.

Zwick, Mark, and Louise Zwick. *The Catholic Worker Movement: Intellectual and Spiritual Origins.* New York: Paulist Press, 2005.

# INDEX

# ABOUT THE AUTHOR

Kevin J. O'Brien is dean of humanities and associate professor of Christian ethics at Pacific Lutheran University. He is the author of *The Ethics of Biodiversity* (Georgetown University Press), coeditor of *Grounding Religion*, and coauthor of *An Introduction to Christian Environmentalism*.